Computers, People, and Thought

Malachy Eaton

Computers, People, and Thought

From Data Mining to Evolutionary Robotics

 Springer

Malachy Eaton
Dept of Computer Science & Information Systems
University of Limerick
Limerick, Ireland

ISBN 978-3-030-55302-9   ISBN 978-3-030-55300-5   (eBook)
https://doi.org/10.1007/978-3-030-55300-5

This Springer imprint is published by the registered company Springer Nature Switzerland AG.
The registered company address is: Gewerbestrasse 11, 6330 Cham, Switzerland

Preface

What This Book Is (Not) About

In this preface, I will initially briefly attempt to answer some observations and potential criticisms that might be made as to the content and style of this book.

On cursory examination of the table of contents of this book, it may be observed that this is not a typical populist AI book. It should be said at the outset that this book was never intended as simply a high-level presentation of the state of the art in AI with all its attendant triumphs and woes. There were a number of books already on this topic at the time of the inception of this project, and there are many more now, quite a lot of which are written by either "non-technical" authors or academics from a different discipline. Rather it was (and is) intended as a focused text, requiring some effort by the reader, but that will hopefully give them a firm foundation of the principles and basics of computing, leading to an appreciation and basic understanding of the recent radical developments in AI and their current and potential impact on humanity.

So, the book is not intended to be "populist", in that sense, but if the careful reader works their way through it will (hopefully) give them some sense of empowerment, in that they will not only have an understanding of the principles underlying the advanced technologies discussed, but some sense also that they can actually *alter* the course of further developments. This is not to say that the book is not meant to be "entertaining" in a certain sense and to hopefully attract a wide audience. I understand that the 1949 book *Giant Brains* by Edmund Berkeley from which this book draws some inspiration was a bestseller in its day.

Another observation could be made that too much emphasis is placed on logic circuitry and Base3 computation in particular—no current AI systems are based on Base3 arithmetic. However, I disagree that an understanding of computation at the most basic level (including logic gates) does not help the reader understand higher level ideas; also, of course, this book is not "just" about AI but also about the ideas and principles underlying computational processes. I accept that ternary computation does not form the basis for any current or planned AI system in the public domain, although I would not be at all surprised if work is progressing "undercover" in this area (particularly in the area of ternary nanoelectromechanical relay-based computers which would be impervious to the effects of nuclear radiation), but again this is not really the point. This book is about principles, and ternary computation is demonstrably more elegant and efficient in many aspects than binary. I also make no apologies for reference to my own work in the area; the ternary ALU I designed and built is to my knowledge the only one of its type in existence, and building a replica of this device would form a very interesting (albeit advanced) project for the technically minded reader.

Another possible criticism might be that too much emphasis is placed on game AI, given that few threats to human liberty and dignity are based on game-playing AI. It is true that few of the threats to human liberty and dignity are based on game-playing AI, but right from the inception of AI, games have been both a major application area and a benchmark for AI systems and as such have contributed hugely to the advanced AI and machine learning algorithms we have today. Also, games are interesting in their own right from a human perspective and form one of the major strands of this book. A related comment/criticism might be that: too much emphasis is placed on so-called Good Old-Fashioned AI (GOFAI). Again, this book is primarily about principles, and GOFAI demonstrates basic principles involved in representation and search, both important issues in intelligent system design. GOFAI is also by no means dead—and still forms the core for many current powerful AI engines.

The material relating to Asimov's laws and the Turing test may be seen as outdated and irrelevant by some, given that no serious scientist or researcher today suggests that we should take either Asimov's laws or the Turing test literally now. However, there is no denying the huge impact that these concepts have made to modern thinking on the development of "safe" robotic and AI systems, and on the benchmarking of the general intelligence of such systems.

Also, the idea of a citizens' convocation may be seen as naive and unrealistic. In response, the idea may appear a little naive—but what is the alternative? This section in fact attempts to address a serious issue—the decision on the power of deployment of modern, potentially damaging technologies is currently in the hands of either governments or large corporations, each with their own agenda, and not necessarily for the benefit of the general citizen.

Finally, it may also be argued that there is no place for a discussion on spirituality and religion in a book such as this. This is a valid perspective, and I urge any reader who feels in any way discomfited by this discussion to skip this short section entirely.

What This Book *Is* About

Following on from my recent focused text *Evolutionary Humanoid Robotics*, I was inspired to write a more general text on AI/Robotics, motivated in part by recent impressive (some might say alarming) developments in these areas.

This may appear at first sight to be a slightly strange book for an unusually wide target audience. I could counter this by the argument that there are perhaps too many books either with too much emphasis on a very particular target audience or books aimed at a very general audience. Not so many texts are aimed at the "middle ground". This book is aimed squarely at this audience: whether it succeeds or fails in this focus is up to you, the reader, to decide.

So this book aims at a clear and lucid exposition of the core ideas in the AI/cybernetics field aimed at both the student and the moderately educated layman alike. We also aim to go beyond this and to look at ways to potentially shape man's relationship with technology into the future for the overall benefit of humankind.

To recap, this book is aimed as

- A text for the moderately intelligent layperson willing to engage with the ideas contained within
- A general introduction/motivational text for students in the areas of computer science and engineering
- An undergraduate-level text for computer science students in the areas of intelligent systems and artificial intelligence in games
- An undergraduate/graduate-level text for students in areas other than computer science introducing them to core concepts in computing and intelligent systems
- An introductory text for researchers/academics in unrelated fields looking for a concise introduction to the core topics covered
- An accessible and authoritative up-to-date text for more advanced students/researchers in the areas of AI/IS/CI/robotics

Acknowledgements
I would like to acknowledge and express my gratitude to those researchers and pioneers who, over the years, through either (some very!) brief encounters or longer acquaintances have helped shape and mould some of the ideas presented in this book. Of course, none of these should be held in any way accountable for any errors or omissions occurring in this modest text—the blame here rests entirely on my own shoulders. They are also in no way responsible for any of my opinions, which some may consider controversial, contained herein.

Also, many thanks to all those who have facilitated my modest research efforts over the years through financial and moral support. I would like to particularly thank Ronan Nugent of Springer for his patience and encouragement through this project, which has taken considerably longer than originally planned. And finally, to my wife Patricia for her forbearance throughout this project to my many evenings and weekends spent scribbling notes or typing at the laptop.

Limerick   Malachy Eaton
February 2020

Contents

Chapter 1
Introduction

Whether we entrust our decisions to machines of metal, or to those machines of flesh and blood which are bureaus and vast laboratories and armies and corporations, we shall never receive the right answers to our questions unless we ask the right questions.
The hour is very late, and the choice of good and evil knocks at our door.
—*Norbert Wiener (1954)*. The Human Use of Human Beings.

1.1 Computers, People, and Thought

In this book we look to three core components of existence today, ourselves (humans), computers—arguably man's most advanced creation and expected by some to replace us someday; and thought, that most ephemeral of phenomena, but without which we would be blissfully unaware of our own existence, and would almost certainly not have created the advanced computers that exist today.

We will also look at synergies between these three entities: between computers and thought, leading to what we commonly call the field of Artificial Intelligence (AI), or Intelligent Systems (IS); between people and thought, leading to questions of consciousness and our own existence as humans; and between computers and people; leading to the recent remarkable advances in the field of humanoid robots.

Finally, we look forward into the (maybe not so distant) future and the "holy grail" of many researchers—the creation of intelligent "conscious" humanoid robots with potentially far superior intellect to ours, yet able to operate perfectly in our native "built-for-human" (BFH) environments, with all of the implications associated with this potential development, that can be—as Wiener says in the quote above—both for good and for evil.

Successive waves of hype have accompanied significant developments in the computing and artificial intelligence fields over the last 70 years or so. Expectations (and fears) have been raised to a feverish pitch, only to fail to be realised, in general, to any significant extent. And, yes, authors in the past have written, as I do, for the informed, reasonably intelligent layperson in order to educate them as to the basic principles and core issues involved from a reasonably authoritative viewpoint. So—

© Springer Nature Switzerland AG 2020
M. Eaton, *Computers, People, and Thought*,
https://doi.org/10.1007/978-3-030-55300-5_1

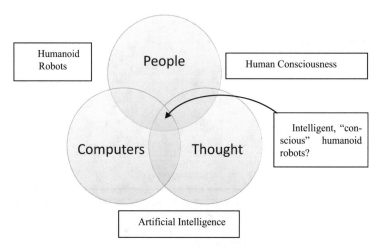

Fig. 1.1 The focus of this text is the confluence of people (us), computers (perhaps our greatest creation), and thought. The field of humanoid robots focuses on the creation of artificial analogues of people with certain intellectual capabilities (C/P). Artificial Intelligence is commonly seen as an attempt to imbue computers with perception and reasoning capabilities (C/T). And consciousness is seen by many as a necessary consequence of the human thinking process (P/T). Finally, looking to the (maybe not so distant) future we may envisage humanoid robots with highly lifelike abilities able to operate with ease in environments built for humans, and able to interact with us in a manner that suggests a certain level of conscious activity (C/P/T)

are things any different this time around? I think so, for reasons that I hope will become clearer in a little while. And this is the reason I have put pen to paper to write this book. See Fig. 1.1 for a pictorial overview of the focus of this book.

1.2   In the Beginning...

In the beginning computers *were* people—in many cases mainly *women*. (And they thought). Then computers became machines, initially mechanical, then electrome-chanical, and now the predominantly electronic wizards that so permeate and (some would say) dominate our everyday existence, certainly in the so-called first world. Very early (non-human) computers could not, except by a very broad stretch of the imagination, be said to have thinking capabilities, but by the late 1940s books with titles such as *Giant Brains, or Machines that Think* (Berkeley 1949), were beginning to appear on bookshelves, to the bemusement of the general public.

Today, with the advent of ever-increasing computer power (both in processing ability and in memory capabilities) in conjunction with clever machine learning algorithms, some eminent scientists and portions of the general public would consider it likely that either we have "thinking machines" in our midst this present day (a minority), or that these entities will come into existence in the near or the not-so-distant future.

Indeed, this belief has become so entrenched in the minds of some eminent scientists and philosophers that a number are calling for restrictions on further work in the areas of machine learning and artificial intelligence until the potential implications of these developments on humankind can be discussed and evaluated more clearly. In fact, for certain controversial areas, such as the granting to autonomous agents (robots) of the right to use lethal force without any human intervention (so-called "killer-robots"), many today are calling for an outright ban on such developments.

Then, there is the notion of computers as people (in physical form). This encompasses the area of humanoid robot development, where the focus is on the creation of robots with broadly human-like features and/or intelligence, Some research in this area is so advanced (especially in Japan) that it can be difficult to distinguish between human and humanoid over short stretches of face-to-face interaction.

A particularly interesting avenue of research in this regard is the *evolutionary robotics* (ER) field, where scientists attempt to recreate, in a computational stratum, the natural evolutionary process that many believe to be the driving force behind the development of the myriad of different life forms extant here on Earth. While the process of artificial evolution may be used to create artificial "creatures" with no Earth-like analogue, many researchers are using ER techniques to create robots that are, in physical form, similar to existing insects or animals, such as centipedes, eels, dogs, etc. The conjecture is that if we can create artificial creatures with the same physical footprint as these natural creatures they will have less difficulty in operating in similar environmental niches to their natural environments (e.g. eels in water, dogs over rough terrain, etc.)—if, indeed, this is the purpose of the research in the first place.

Of course, another reason we might wish to design/evolve artificial creatures in the general form of insects or animals might be to study more closely the behaviour of these same creatures in the natural world, perhaps performing experiments that would be difficult or unethical if performed on real animals.

This brings us nicely to the field of *evolutionary humanoid robotics* (EHR). While in the ER field the goal is to use an artificial evolutionary process to create the body and/or "brain" of a robot which may (or may not) resemble in its gross behaviour and/or appearance a living creature, the focus in evolutionary humanoid robotics is firmly on the creation of human-like appearance and/or behaviour through a process of artificial evolution.

The extent of the likeness of the evolved humanoid to a real human may, of course, vary—from *replicant/android* level where the robot is virtually identical to a human in every physical and behavioural aspect, to what we will term *built-for-human* (BFH), where the main requirement on the robot is that it is capable of operating in a variety of environments designed for humans; however it is not required to look like a human or to behave in a particularly human-like fashion (Eaton 2015).

1.3 Full Circle?

And so, in a sense, we have come full circle. For, if it will be possible (as computer scientists and roboticists are increasingly arguing) to move towards producing a robot at android/replicant level (we use these two terms interchangeably, however the replicant level is considered to be more rigorous in terms of human appearance and behaviour), through evolutionary or other means, many consider that it will become increasingly difficult not to imbue these robots with the apparent ability to feel emotions, and, yes, even to think.

So, we have traversed the circle, from a computer as a person with undoubted thinking abilities to an artificial computer/robot, which we may term an *advanced artificially intelligent entity*, or A^2IE for short, possibly taking the form of a person, and which we may construe as having the ability to think, and even to possess consciousness (however we define this). However, we might well ask whether this is a route we wish to take—the development of artificial human-like robots in form and behaviour virtually identical to humans, with all of the philosophical, societal, and ethical implications involved.

There is one thing that we can be sure of, and that history has taught us. Progress in science and technology will continue—it is up to us all now to decide whether we wish to put in place caveats and restrictions on the use of these advancements, as is the case with other disruptive technologies, or to allow their unfettered development, with all that this might imply.

1.4 Amazing Technologies (What Has Changed)

Don't get me wrong. I like my technology. Especially when it's not being used to track me, monitor me, sell me things, influence my behaviour in sneaky ways, and all of the myriad of other not-so-pleasant things that modern technology is increasingly being used for.

Now we might say that we are in an era where AI is poised to be the "next big thing"—but then again people also said this in 1997 when world chess champion Garry Kasparov was beaten by the chess-playing computer Deep Blue, and in the 1980s when expert systems were all the rage, and even back in the late 1940s/early 1950s when books with titles such as *Giant Brains or Machines that Think* (Berkeley 1949) were sold to enquiring (and possibly fearful) readers.

So, after all of these years, what has changed really? Well, one thing that has certainly changed is the ability of AI techniques to have a direct impact on everyday human lives. The defeat in 1997 of Garry Kasparov by Deep Blue was definitely an important event in both academic and wider public circles, but by no stretch of the imagination could it be said to have had a major impact on the lives of everyday people going about their normal lives.

However, recent developments in the area of image recognition, in particular for example in the area of human face recognition using so-called "deep learning" techniques cannot be said to be of such benign character. For example, at the present time, current technology allows for better than human-level face recognition in video images, with all that this implies. Also, allied technology allows for the prediction of intimate and private human characteristics such as religion and sexual orientation from simple Facebook "likes" (Wang and Kosinski 2018). Also, of course, there is the imminent introduction of autonomous vehicles on our roads. In the US alone there are estimated to be in the region of 2.2 million truck drivers whose jobs are potentially threatened by this development.

1.5   Past Versus Future AI Advances

So, advances in the past, while undoubtedly significant (no-one can sensibly say that the defeat of Garry Kasparov by Deep Blue in 1997 was not an astonishing achievement in human creativity and ingenuity), have not, however, had huge impact on human society, in general. But coming advancements will. Once an advanced artificially intelligent humanoid robot has been developed, in the same form as a human, able to operate in many circumstances as a human can, and, in some cases, where a human cannot, the whole picture changes radically.

It is my belief that such developments are taking place and are currently at a reasonably advanced level such that this eventuality (and it will be the case) is not many decades away, but far closer than this. I predict that by the year 2030 advancements in humanoid robot and AI/IS technologies will be such that we will have humanoid robots with the ability to co-work with their human counterparts in a wide range of occupations. From there—where to next?

1.6   The Accelerating Pace of AI Developments: A Case for Restraint?

The whole AI field is shrinking at a remarkably rapid and accelerating pace if one views the AI domain as those areas in which humans outperform artificially intelligent entities (AIEs). This is why it is more important now than ever for the average citizen to become informed of its astonishing potential for good, and it's equally astonishing (some would say more so) potential for harm.

As an example of the "infiltration" of AI technologies into day-to-day human activities nowadays, watching people with their heads down over their smartphones on the street, in cafes and restaurants, oblivious to the world around them, it is hard to escape the impression that, perhaps, just perhaps, it is not them, but their technology, that is the master. It is just an impression maybe, but a powerful one nonetheless. Also, talk of possible "Terminator"-style robots in the future ignores the fact that we, today, have the capability of handing over decisions on the use (or not) of lethal force to machines, [if, indeed, this has not already happened]. Also, for example, there is the fact that your shiny new "smart" television set may, in fact, be spying on you.

However—used wisely—these new technologies could result in improved (indeed previously unheard of) living standards for us all, with significantly reduced working hours, and the possibility of regular "sabbaticals" for all workers, with the vast majority of unpleasant labour being removed from the remit of human workers. In this book we squarely address those concerns of members of the general public and of the academic community (both those affiliated with the AI field, and those not) about the potential damaging effects of current and future AI technologies.

1.7   Recent Book Publications in the Area of Advanced AI Systems and Future Technological "Progress"

This book is a little different to many of the texts on this general topic that you may have seen on your bookstore shelves in recent months or years. It is not written by a journalist or a "freelancer" but rather by a computer scientist with over 25 years' experience in industry and in academia in the AI and robotics fields. In reaction to this you might say—well this will probably be a fairly turgid text appealing only to the hard-core practitioner, or perhaps a fairly dry educational text. But this is certainly not the intention of this book. Quite a number of books have been published addressing some of the issues raised in the latter parts of this text since I first conceived of this project—this could be seen as making my life easier, or more complicated. I suppose the reality was, in hindsight, both.

An analogy could be drawn here, with the widespread availability of Internet access, and the ubiquitous use of powerful search engines such as Google. On one hand it makes the writer's task easier by providing, among other things, a powerful reference tool for almost any topic. On the other hand, because of the ready availability of such tools to the general public (though they may well lack such ready access to many learned articles and writings made available to university denizens), this makes for a generally more informed and critical public. Gone are the days when the ordinary citizen presents him or herself at the door of their GP with some minor ailment or other, without first having performed a thorough self-diagnosis via the Internet (while probably scaring themselves silly in the process!).

In one sense it is refreshing to think that so many (mostly) talented writers have taken an avid interest in the topic to the extent that they are prepared to spend the long and arduous hours necessary for their product. What is perhaps less pleasing is having to sift through this recent array of work seeking out the interesting and well-researched items from the perhaps not so well so. What is particularly striking this time around is the number of "non-technically-qualified" (to put it nicely) writers generating a plethora of articles and books on the subject. This, of course, is not to say that academics (myself included!) do not, at times come out with writing may that may not reach the highest levels.

Relatively few writers (with some notable exceptions) have supplied so much technical detail for a popular book to back up their musings and convictions. In this text I hope to supply the reasonably intelligent reader, from whatever background, with sufficient detail to become relatively "dangerous" in their convictions, such that they are in possession of sufficient basic technical knowledge, based on their reading, to argue their case more clearly.

However, in the course of my research for this project I came across several books either just published or only published recently (the last few of years), that, on initial inspection, led me to think that perhaps I need not continue with my own project—maybe the work had already been done for me. Such has been, and is, the level of interest and activity that the general topic of the fabrication of intelligent "thinking" agents (whether embodied or otherwise) is currently provoking. On closer inspection of these texts however, I concluded that my own approach to the subject matter was, if not completely unique, at least sufficiently different to warrant a separate text. Whether this conclusion was justified or not is left up to you, gentle reader, to decide.

So, a veritable explosion of relevant texts have emerged in the last few years, including Nick Bostrom's pioneering *Superintelligence: Paths, Dangers, Strategies* (2014), Martin Ford's *The Rise of the Robots: Technology and the Threat of Mass Unemployment* (2015), *The Glass Cage Automation and Us* by Nicholas Carr (2014), *The Master Algorithm: How the Quest for the Ultimate Learning Machine Will Remake Our World* by Pedro Domingos (2015), *Rise of the Machines: The Lost History of Cybernetics* by Thomas Rid (2016), John Brockman's *What to think about machines that think: today's leading thinkers on the Age of Machine Intelligence* (2015), *The Cyber Effect: A Pioneering Cyberpsychologist Explains how Human Behavior Changes Online* by Mary Aiken (2017), *Social Machines: The Coming Collision of Artificial Intelligence. Social Networking, and Humanity* by James Hendler and Alice Mulvehill (2016), *Heartificial Intelligence: Embracing Our Humanity to Maximize Machines* by John Havens (2016), Luke Dormehl's *Thinking Machines: The Inside Story of Artificial Intelligence and Our race to Build the Future* (2016), *Technocreep: The surrender of Privacy and the Capitalization of Intimacy* by Thomas Keenan (2014), Sherry Turkle's *Reclaiming Conversation: The Power of Talk in a Digital Age* (2015), Lipson and Kurman's *Driverless: Intelligent Cars and the Road Ahead* (2016), Max Tegmark's excellent *Life 3.0: Being Human in the Age of Artificial Intelligence* (2017), *Towards a Code of Ethics for Artificial Intelligence* by Patricia Boddington (2017), Steven Pinker's *Enlightenment Now: The Case for Reason, Science Humanism and Progress* (2018), the excellent *The Digital Ape: how to live (in peace) with smart machines* (2019) by Nigel Shadbolt and Roger Hampson, and most recently *Responsible Artificial Intelligence* by Virginia Dignum (2019).

1.8 "Thinking Machines" Revisited

When I saw the title of a recent book on AI entitled *Thinking Machines* (Dormehl 2016), I thought "I'm sure I've come across that title before"—and indeed I had... several times. One of the earliest books with a similar title was Edmund Berkeley's pioneering 1949 bestseller *Giant Brains, or Machines that Think* (Berkeley 1949). But there have been many texts with similar titles published in the interim period. On my own bookshelves are the classic 1963 collection of papers *Computers and Thought* (Feigenbaum and Feldman 1963), Igor Aleksander and Piers Burnett's *Thinking Machines: the Search for Artificial Intelligence* (Aleksander and Burnett 1987), Irving Adler's 1951 text *Thinking Machines: a Layman's Introduction to Logic, Boolean Algebra and Computers* (Adler 1961), McCorduck's (1979) *Machines Who Think*, and John Brockman's more recent *What to Think About Machines that Think* (Brockman 2015).

And I am sure there are many more. Clearly the topic of "Thinking Machines", whatever exactly this might mean, and how it might manifest itself in our ordinary day-to-day existences, has increasingly exercised the human imagination, especially since the prospect of such an eventuality became a distinct and realistic possibility in the eyes of some, from the late 1940s onward.

As mentioned, a founding seminal text in the area of intelligent systems was that published in 1963 and entitled *Computers and Thought*. This book comprised an anthology of many of the most important writings in the AI/IS field up until that date. In the current text we attempt to put *people* back into the loop with all of our innate human physical, intellectual, and emotional idiosyncrasies. Why? Because people *matter* (see Fig. 1.2).

Fig. 1.2 Bringing humans more firmly into the picture

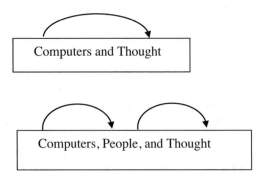

Computers and Thought

Computers, People, and Thought

Core issues:
- bridging the "gap" between computers and thought
- bringing the human element firmly into the picture
- the importance of embodiment

We will focus more on important principles, rather than on historical aspects or individual personalities. Of course, some level of detail on history and personalities is warranted when certain individuals or certain historical events have had a major shaping influence on the field; however, given the compact nature of this book detailed discussion of these aspects is left to other texts, of which there are several.

1.9 Questions, Answers, and Goals: The Focus of This Book

This book aims at a clear and lucid exposition of the foundations of computing and the core issues of artificial intelligence at the present time and into the future. The main purpose of this book will be to inform, rather than to act as a historical document.

This is not a particularly long book, nor was it intended as such. Long books have a tendency to remain unread, especially if they are perceived to contain material that may be construed as particularly intellectual or challenging. Or, at least, vast swathes of content may remain untouched—thus the reader may possibly miss the core focus and thesis of the book (assuming, of course, that these components do exist in the first place). Of course, this book does not necessarily provide all of the answers to those seeking illumination; however, it does provide significant information in a form that should be accessible to the moderately educated reader who is prepared to invest a little effort in its assimilation.

This book does not have a single specific focus. Its aim is to educate (and also in fairness, to an extent, to entertain), so that hopefully by the end the careful reader will be more aware of the computer machinery that we have in our midst at the present time, and how it may interact with us into the future, and by having this enhanced awareness will be able to make better-informed decisions as to what technologies will be of benefit to humankind into the future, and what, perhaps, will not.

This book also has the twin unifying themes of games—the games people, computers, and robots play—and evolutionary systems, based on Darwin's "survival of the fittest" (Darwin 1859) and Mendelian genetics (Mendel 1866). It also presents a clear message—machines and technology have an important role to play in enhancing human lives, but not at the expense of the loss of core elements of human existence and core human values. This book is also intended to be an accessible perspective on the core principles underlying both the challenges and the many opportunities offered by the current and future explosion of "AI-type" technologies and other advanced expertise sets. Forget about the hype about "cyborgs", or humans transferring their consciousness into machines. There are real, more pressing, concerns of relevance to every man, woman, and child on the planet today.

This book does not, however, reach one simple conclusion. Like almost all advanced technologies there are many uses both for good and for bad. At least for the medium-term future the main negative aspects of these advanced technologies appear to be in the hands of humans rather than machines. We mentioned the rapid erosion of privacy accelerated by these technologies—this issue can only become more pronounced in the future unless steps are taken in terms of laws, etc. in this regard. The whole area of the employment prospects of humans versus machines is also likely to come to the fore in the not-too-distant future.

Other, more exotic dangers, such as the development of autonomous "super-brains" harbouring dreams of world domination (or worse), are still firmly in the realms of science fiction writers, and will remain so for some time into the future. But who is to say what advances might be made, and what the potential consequences of these advances might be for humanity? As recently as 2015 the majority of AI scientists would have predicted that the advent of a computer taking on, and beating, the world Go champion was at least a decade away. Then along came AlphaGo—and the rest is history.

However, the major focus is not to cause fear or a sense of paralysis in the face of future technologies. We, now, all of us, at the present have the power to change and to alter in subtle and not so subtle fashions the trends of future technologies. However, we need to be aware that, as these technologies become more entrenched and accepted by the mass of humanity it may become difficult, if not impossible, to roll back the clock. In a nutshell, time is of the essence.

There are a lot of books out there about robots and technology and how humans are increasingly becoming redundant in the workplace and irrelevant in the greater scheme of things. This is not one of these books. This book is about spreading a message of hope, not of fear. We are humans, we make the rules, and we can make those choices that need to be made. Technology can be a double-edged sword, with both favourable and unfavourable consequences. The more powerful the technology, the sharper the edges. And very powerful technologies indeed are either currently available, or under rapid development as we speak.

And, above all, let us not forget the potential enormous benefits to humanity of these technologies, used in a wise and thoughtful fashion. Since the dawn of the Stone Age mankind has fashioned implements and devices to augment human competence and performance, in one area or another. The stone axe facilitated the defence of Stone Age man from his predators in a way that bare hands did not. But the stone axe was not a replacement for bare hands; this may prove also to be the case with modern technological advances.

What is it that we, as humans, want for our own future and for future generations? Technology unbridled, with all that encompasses, or a more controlled and compassionate, yet still advancing technological frontier geared toward the betterment and advancement of humankind as a whole? I, for one, know which side of the fence I stand on. The title of this text is *Computers, People, and Thought*. And there is plenty for us people to think about in relation to computers (and associated advanced technologies). But let us go further—let us put our thoughts into action.

1.9.1 Difficulty Level of Some of the Material Contained Herein

While we aim at a clear and lucid approach to the subjects contained herein, not all of the subject matter will necessarily be easily comprehended in a single sitting because of the inherent difficulty level of some of the subject matter; especially if the reader has no previous exposure to this material. However, persevere and hopefully you will be rewarded with a reasonably deep insight into some topics that will very likely have a fundamental impact on our lives, now and into the future.

1.9.2 Book Is About Principles, Not Players

To be clear, this book is not about me, nor intended to describe in any level of detail any of the main protagonists of the computational principles and ideas contained herein. There are many other texts ostensibly suited to this task. This is not to trivialise in any way the achievements of these people—just to acknowledge that the basic principles of computation and artificial intelligence, like, in a sense, the prime numbers and the whole field of number theory, were there in the first place—just waiting to be discovered. However, certain figures have made such an important contribution to the field that it only seems proper to acknowledge their influence. In Chap. 5 we briefly sketch the lives and contributions of nine such people.

1.9.3 Main Focus of Algorithms in This Book

Because of the focus and also the relative brevity of this book we will concentrate mainly on the so-called *bio-inspired* paradigms; that is, those learning and other paradigms that take their inspiration either directly or indirectly from biological organisms. These algorithms have a compelling "proof-of-principle" background. Examples of these paradigms include *evolutionary algorithms* (EAs), broadly modelled on the processes of natural evolution, and *artificial neural networks* (ANNs), broadly inspired by the neural structures and presumed functioning of the brains of humans and other species.

A slightly more general category of systems—the so-called *nature-inspired* algorithms—are inspired by any aspect of learning or development in the natural world, not exclusively (but obviously including) the biological realm. The field of *simulated annealing* belongs, for example, to this broader category. This algorithm is inspired by the annealing process in metals, which when heated up to a high heat, and then allowed to cool, gradually tend towards optimally compact structures. There is an even broader category of "metaphor-inspired" algorithms, which may also derive inspiration from human behaviours or other extant phenomena. We will take a slightly more critical and nuanced look at these algorithms in Chap. 5.

1.9.4   A Note on the Book Subtitle: "From Data Mining to Evolutionary Robotics"

The subtitle of this book *From Data Mining to Evolutionary Robotics* in a sense follows my own career progression, moving from the field of what would now be called data analytics/data mining, through work in the areas of data communications and compiler design to my current main area of interest—*evolutionary robotics* (ER), or, more specifically, *evolutionary humanoid robotics* (EHR); that is the application of evolutionary principles to the design of humanoid robots. There is a significant possibility that either (or, indeed both) of these domains will play a major role in the development of future advanced intelligences.

1.10   Recap: Overall Book Layout, and Intended Audience

The major focus of this book is on the moderately intelligent and educated lay person, ignorant of the issues discussed here, but aware of future impending technologies, and who also understands the importance of learning more about their fundamental workings and their potential future implications. These readers may perceive the need to educate themselves in this rapidly advancing field, which is now seen as having a dramatic impact on many aspects of everyday life, and will continue to do so increasingly into the future.

This book can also be seen as a stepping stone for researchers in a diverse range of AI/robotics-related topics to extend their expertise into some other related area discussed in this book. However, we aim to go further in this book, with a brief introduction to the field of computation, in keeping with the early "AI" texts, but broadening our perspective to include a discussion of non-binary computation, in particular the ternary system, described by Donald Knuth in his influential book series *The Art of Computer Programming* as "Probably the prettiest number system of all" (Knuth 1981).

Another goal is to serve as an ancillary textbook for students in the Computing/ AI/Intelligent Systems fields. Although this book will touch on subjects that may be considered difficult, such as human consciousness and cognition, it will do so in layman's terms as much as possible, with the aim being to clarify rather than to obfuscate core ideas. In general, we aim this to be the tone of this book.

Following this introductory discussion, the next part will explore separately in some detail each of the three main components of this text: Computers, People, and Thought. Because of the focus of this text the majority of Part I will be devoted to the mechanics of computation. Part II then looks to the synergies between these three main components in more detail, while Part III looks at the confluence of all these components. Finally, in Part IV we look at potential future scenarios for humanity and "mechanisty".

Those readers interested in gaining a deeper insight into some of the more esoteric topics covered in this book should refer to the supplied appendices; however, these are not required reading for the general reader interested in gaining an overall comprehension of the topics covered. In addition, this book is scattered with examples, some worked out thoroughly, to give the reader a flavour of the practical aspects of many of the topics covered, and to encourage further experimentation by the curious reader.

Part I
The Components

Chapter 2
Computers

The Analytical Engine consists of two parts: —

1st. The store in which all the variables to be operated upon, as well as all those quantities which have arisen from the result of other operations, are placed.

...
2nd. The mill into which the quantities about to be operated upon are always brought.

There are therefore two sets of cards, the first to direct the nature of the operations to be performed—these are called operation cards: the other to direct the particular variables on which those cards are required to operate—these latter are called variable cards.

...
The Analytical Engine is therefore a machine of the most general nature. Whatever formula it is required to develop, the law of its development must be communicated to it by two sets of cards. When these have been placed, the engine is special for that particular formula.

—*Charles Babbage (1864).* Passages from the Life of a Philosopher.

2.1 Introduction to Computers and Computation

A little over 160 years ago an Englishman, the first professor of mathematics at Queen's College in Cork, Ireland (now University College Cork, and part of the National University of Ireland), George Boole, published what was to become one of the seminal works in science and laid many of the foundations for modern computing. The title of his treatise was *An Investigation of the Laws of Thought*, and this book laid the foundations for what we now call Boolean algebra, one of the cornerstones of the modern binary digital computer.

The term "computer" can mean one of several different things. In the early 1940s the term was commonly used to refer to people, sometimes called *girls* (because they generally were women), who performed computational tasks at a relatively high speed, with the aid of mechanical devices. Today the term is used almost exclusively to refer to binary, digital electronic machines capable of performing high-speed operations on data. While we will extend this rather narrow view of computers and computation in a little while it is probably useful at this stage to come up with a simple one-sentence overview of a computing device as adapted from C.S French (1980)

> A device operating under the control of a stored programme(s) automatically accepting and processing data.

Note here the use of the original (and now considered rather old-fashioned) "programme" rather than "program"—perhaps a slightly more evocative term. Modern digital computers are generally mainly electronic in nature, while possibly incorporating electromechanical components such as hard disk drives. However, in the past computers were constructed on mechanical principles (cogs, pulleys, etc.), and also by using mainly electromechanical components (various types of relays, etc.). It is even possible to construct computational devices based on fluidic principles. But the basic principles of digital computation remain the same. Information is represented as a number of discrete states (just two states in the case of a binary computer) which is then manipulated according to a number of rules, or instructions. In a *stored-program* digital computer the set of rules or instructions is stored in the computer itself, prior to execution. Early computers did not employ this concept, instructions had to be fed in separately from the data—but virtually all modern computers employ the stored-program concept. Note that here we use the terms *data* and *information* almost interchangeably, distinctions are made in some literature—generally information being described as a more refined form of data, or data following processing of some kind, however these distinctions can prove confusing to the uninitiated.

2.1.1 Computer Power as the Epitome of Modern Technological Progress

The title of a popular, and widely referenced, book on the topic of Artificial Intelligence is *Artificial Intelligence: Structures and Strategies for Complex Problem Solving* (Luger 2009). But—isn't that just about writing programs, you might ask (albeit programs for "complex problems", whatever those might entail)? Well: yes indeed—but only up to a point. This is why we need to start this part with a short exposition about the very nature of digital computation itself.

Computation, and computer power, to a large extent nowadays epitomise modern man's technological progress and prowess. They are, in a sense, today, humankind's ultimate intellectual achievement. Others would look to areas such as nanotechnology and biotechnology, or perhaps nuclear technologies or space exploration as other major markers of human technological and intellectual prowess. But all of these other advances have at their core one enabling technology—computers, and computer power. Interestingly, it turns out that modern computer games are excellent showcases for advances in computer power and, in many cases, drive further advances in computing technology—certainly in the popular consumer market. So, while there are other huge technological advances in areas such as biotechnology and nanotechnology, none of these appear to have the same capacity for direct impact on humanity and human "evolution" as computers and their potential future progeny, *Advanced Artificially Intelligent Entities* (A^2IEs).

2.1.2   Computers: Switches Driving Switches

As part of my Computer Organisation course for first-year undergraduate computer science students I demonstrate part of the working of a full adder (a core component of the computer's central processing unit (CPU)) using a number of interconnected switches to generate the correct responses, displayed as lit or unlit bulbs, from the device as shown in Fig. 2.5. I cover over the interconnections and ask students how they think the logic is implemented in hardware. When I reveal the inner workings of the device the response is "but that's just a bunch of switches".

Exactly. Because that is what I am demonstrating to these students—a number of multi-pole switches connected together in a clever fashion. And that fact is precisely what is at the core of a digital computer—be it binary, ternary, or whatever—a (very large) bunch of switches interconnected in a particularly ingenious fashion. Nothing more, nothing less. This poses an interesting question, core to this book, which we will address in more detail later—as humans, do we really wish to be reduced to a synchronised bunch of switches? Whatever the brain is (if we consider the brain to be the seat of thought), it is most certainly not a serial synchronous digital binary computer, which broadly characterises the vast majority of computing devices on the planet today.

2.1.3   Relays, Vacuum Tubes, and Transistors as Switches

Relays are essentially just electromechanical switches. This just means that rather than manually moving a lever or knob, the switching action is activated by electrical means. Vacuum tubes and transistors are also, on one level, just switches in electronic form. While their behaviour is more sophisticated than that of simple switches, in their use in digital circuitry (as opposed to analogue) it is just their ability to act as very fast switches that we are interested in. And, of course, it is the combination of huge numbers of transistors and other associated electronic components in a highly miniaturised form that comprises the integrated circuits that drive today's powerful computers.

2.2 How Computers Represent Information About the Real World

2.2.1 Data in Its Many Guises

Data may take one (or more) of many different forms. It may take the form of measurements of real-world conditions, such as temperature, visual data, sound, etc. These data will generally be continuous in nature (analogues of actual conditions, or, so-called, *analogue* data), and, as such must be converted to discrete, digital form for processing by a digital computer. Other data will be discrete in their essence, such as integer data, letters, and other symbolic data. The only conversion required for these data is conversion to the number of discrete states used by the computer; in the case of a binary computer this is two. These two states in a binary computer are generally referred to by the digits 0 and 1. While almost all modern computers operate using binary, it is also possible to have a computer representing data using 10 discrete states, corresponding to our decimal numbering system, or, say three states—a ternary system. Both of these systems have been used in the past for the representation and processing of data. The 10-state system has the advantage of allowing for the input of decimal data directly, without a conversion step from decimal to binary. This is an especially useful feature if the only type of processing done on your computer involves numeric data—so-called *number crunching*; what is a little problematic here is the representation and processing of these 10 separate states internally in the computer.

2.2.1.1 Numbers

Of course, the representation of data in binary form brings its own difficulties. However the difficulties that may be encountered in this regard can in many ways be seen to be more than made up for by the relative simplicity of the circuitry required to perform computations in binary rather than in decimal (remember the difficulty you had in learning the decimal multiplication table...). Multiplication and addition tables are considerably simplified in binary; however, binary is *not* the simplest system. The simplest system is sometimes called the cardinal system, or unary system, where there is just a single symbol, repeated as many times as the size of the number you wish to represent. We could use fingers, pebbles, or electrical pulses to represent this single symbol. A major problem with this system, of course, is that because of its simplicity it is not really practical for any but the simplest calculations. Adding 3 plus 2 on your fingers is fine—adding 30 plus 20 is not.

So, we come to the so-called "place-value" or "positional" system. In this system the *position* of each symbol in the overall number has a direct impact on the relative weight given to that number. The overall number of symbols employed denotes the base (sometimes called the *radix*) of the number system, so for the decimal system (10 symbols overall) the number 352 indicates that we have three units of 100 items

Table 2.1 Binary to decimal conversion

1	0	1	0	0	0	0	1	0	1	
$\times 2^9$	$\times 2^8$	$\times 2^7$	$\times 2^6$	$\times 2^5$	$\times 2^4$	$\times 2^3$	$\times 2^2$	$\times 2^1$	$\times 2^0$	
512+	0+	128+	0+	0+	0+	0+	4+	0+	1	$=645_{10}$

Table 2.2 Ternary to decimal conversion

2	1	2	2	2	0	
$\times 3^5$	$\times 3^4$	$\times 3^3$	$\times 3^2$	$\times 3^1$	$\times 3^0$	
486+	81+	54+	18+	6+	0	$=645_{10}$

(300), five units of 10 items (50), plus two individual units. Written in a slightly more mathematical format $352 = 3 \times 10^2 + 5 \times 10^1 + 2$. This convenient and compact method of representing numbers is, by far, the most commonly used in the world today.

A major exception to the use of "pure" place-value systems worldwide is the Chinese numeral system, in which numbers are written in a similar fashion to how they might be spoken in the English language. For example, the number 243 would be written in Mandarin as 二百四十三, literally two hundred (二百) and forty (four 10s) (四十) three (三). However, the "Arabic" numerals (0–9) together with standard positional notation are also widely used in China. In fact, it may be argued the Chinese method of writing numbers should not be considered as a numeral system per se. in the same way that the English phrase "Eight thousand" is not. Let us now look briefly at the representation of numbers using a few different number bases. Taking the binary number 1010000101_2 (we use a subscript to indicate the radix if this is not clear from the context) we have the binary to decimal conversion process, as shown in Table 2.1.

Table 2.2 demonstrates the conversion to decimal of the corresponding ternary (base 3) number 212220_3:

Immediately, we see that the ternary representation is more compact than the binary representation—fewer *trits* (*ter*nary dig*its*) than *bits* (*bi*nary dig*its*) are required. This is as we would expect—however we must remember that each individual trit contains higher information content than each bit and so will be more complex to represent in a computational substrate. This brings to mind an interesting question that we will come back to shortly: is there some way of calculating the most efficient overall representational mechanism for data in the computer?

Looking at one further number base, base 16, or *hexadecimal,* for the same number, we have as in Table 2.3:

This, again, gives us a more compact representation, as we might expect. An issue, of course, immediately arises with hexadecimal notation, one that we have not encountered with either binary or ternary. Because the radix is greater than ten, we need to supply more symbols in order to represent the additional six characters required. By common convention the first six characters of the alphabet, A...F, are used (or, less commonly, the lower-case letters a...f) in order to represent decimal

Table 2.3 Hexadecimal to decimal conversion

2	8	5	
$\times 16^2$	$\times 16^1$	$\times 16^0$	
512+	128+	5	$=645_{10}$

Table 2.4 Hexadecimal to binary conversion

A	B	B	A	
$\times 16^3$	$\times 16^2$	$\times 16^1$	$\times 16^0$	
40,960+	2816+	176+	10	$=43,962_{10}$

Table 2.5 Decimal, binary and hexadecimal equivalents

Decimal	Binary	Hex
0	0000	0
1	0001	1
2	0010	2
3	0011	3
4	0100	4
5	0101	5
6	0110	6
7	0111	7
8	1000	8
9	1001	9
10	1010	A
11	1011	B
12	1100	C
13	1101	D
14	1110	E
15	1111	F

numbers in the range 10–15. For example, ABBA (the hexadecimal number, not the pop group) converts to decimal 43,962 thus we have as in Table 2.4:

But why our interest in hexadecimal at this point, rather than, for example, *duodecimal* (base 12) which is also a useful radix, and which has had application in monetary and other systems. In fact, in general the greater the number of divisors, the greater the utility of a radix—for example 12 has the divisors 2, 3, 4, and 6; 10 by contrast has just 2 and 5. Our interest in base 16 stems from the fact that it is an exact power of 2, allowing for a concise and convenient representation of binary numbers. And, as virtually all modern digital computers employ the binary system, this can be no bad thing. Here in Table 2.5 is a conversion table for the first 16 integers, 0–15, in decimal, binary, and hexadecimal formats.

Use of this table allows for an easy and rapid transition between data in binary format and its equivalent hexadecimal representation because, as we can see from the above table, there is an exact equivalence between any group of four bits and its *hex* (short for hexadecimal) equivalent. To perform this conversion we simply arrange the bits in the binary number in groups of four (starting from the right of the binary number), and assign to each grouping the equivalent hex digit: to go in the

Table 2.6 Hexadecimal equivalent of a long binary number

Binary	1000	1000	1000	0011	0101	0101	0000	1010
Hex	8	8	8	3	5	5	0	A

opposite direction (hex to binary) just replace each hex digit by its four-bit binary equivalent, as given by the table above. The corresponding conversion from binary to decimal and vice versa is a far more complex task involving in one direction repeated division of the decimal number by 2 and noting the remainders generated, and in the other direction the procedure outlined earlier of adding together various powers of 2.

Modern computers typically have *word sizes* (that is, the typical size of the grouping of bits manipulated by the computer at one time) of 32 or 64 bits. Converting between a 32- or 64-bit quantity and decimal is difficult and cumbersome. It is also virtually impossible to visualize the impact of any individual bit in a particular word, and this is something that may be of interest to us as, for example, a single bit setting might represent a valve being in an open or a shut position in the context of a chemical plant or a nuclear power station.

With hex these tasks of visualisation and conversion become relatively trivial. For example, take the binary number 10001000100000110101010100001010. At first sight this is just a jumbled mass of bits. Its decimal equivalent 81,737,317 doesn't tell us very much either. But let's look at its hex equivalent: This is given in Table 2.6.

So $10001000100000110101010100001010_2 = 8883550A_{16}$. This is a more compact representation of the binary number than its decimal equivalent, and a lot easier to generate, and we can immediately deduce aspects of the binary number that are not obvious from its decimal representation. For example, we can immediately see that all of the bits in positions 4–7 are zero. (The general convention is that we start labelling bit positions at the rightmost bit, starting at bit 0.) This fact is not at all obvious from the decimal equivalent, and this might be of interest to us if, for example, that each of the bits in these four locations represent the on or off positions of a lever or a valve in a nuclear power plant.

2.2.1.2 Letters and Other Symbols

So, we can represent decimal (integer) data quite straightforwardly in a binary digital computer, if this is our wish. But what about non-numeric data, symbols such as *, %, £, etc. together with the letters (upper and lower case) of the alphabet? As we start this discussion it is important not to confuse the capital letters A–F as used to represent the hex equivalents of the decimal numbers 10–15 with the representation of the *characters* A–F. Remember we could have used any other symbols to represent these numeric values. The important thing here is that we establish a 1-1 correspondence between a sequence of binary digits and a particular character that we wish to represent. The other important thing is that the particular method that we

Table 2.7 Table of powers of 2

Bits	1	2	3	4	5	6	7	8	9	10
No. of variants	2	4	8	16	32	64	128	256	512	1024

use to encode these characters is also known to anybody else we might wish to have dealings with our data: for this reason it makes sense to employ a standard conversion method between sequences of bits and particular characters. A commonly used conversion table is the *American Standard Code for Information Interchange* (ASCII). This code does exactly what its name implies—it provides for a standardised conversion between a sequence of bits and a particular character. ASCII is a 7-bit code (generally expanded to 8 bits, as 8 is a power of 2 and a more efficient grouping to deal with in a binary computer). Let us see how we can use a code such as ASCII to represent letters and other symbols.

With a single bit we can distinguish between two different quantities—0 and 1; A and B; + and *, etc. With two bits, we can distinguish between four, etc. It is useful to build up a table of powers of 2—here in Table 2.7 is one up to 2^{10}.

Looking at these figures, a couple of interesting issues become clear. Firstly, we can clearly see the number of bits required to encode different types of data. Secondly, there is the fortuitous connection that 10 bits allow us to encode 1024 different quantities (approximately 1000): hence the generally used equivalence 1K = 1024 bits.

The main point of the use of a code such as ASCII in order to represent symbols is to have a particular scheme for assigning bit-patterns to symbols, and to stick with it. So, for example, the ASCII code for the upper-case letter "D" is 44 (hex). If we transmit this bit pattern (01000100) to anybody using another computer anywhere in the world, once they know that we are using the ASCII code they can immediately decode this bit pattern to obtain the underlying symbol.

By judicious design of these coding systems we can also come up with programming shortcuts. For example, in ASCII all of the upper-case letters are ordered in sequence, starting with "A" at 41_{16}. Similarly, for the lower-case letters, but this time starting at 61_{16}. So, the difference between the code for an upper-case letter and its lower-case equivalent is 20_{16} ($61_{16}-41_{16}$) or 00100000_2. By adding this value to the ASCII code for each uppercase letter we can directly convert it into its lower-case equivalent. This can be done by simply setting the third bit from the left equal to 1.

In a similar fashion we can represent other symbols such as punctuation marks and mathematical symbols. ASCII also allows for the encoding of certain *control* characters, to allow for the governing of the flow of information.

Of course, to represent languages such as Japanese and Chinese, plain ASCII is nowhere near adequate. So, in more recent times more advanced coding schemes such as Unicode and UTF-8 have been developed, which allows for the representation of over a million unique symbols. These schemes are generally designed to be backwards compatible with ASCII.

Fig. 2.1 A prehistoric cave
drawing?

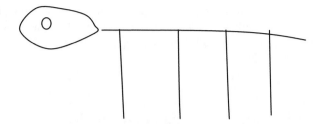

Fig. 2.2 A digitised
artwork

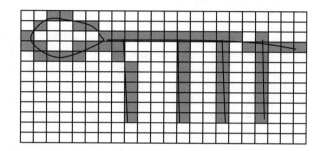

2.2.1.3   Pictures and Sound: Real-World Data

As well as numbers and letters and symbols such as the capital letters A–Z, and the
lower-case letters a–z, it is also possible to represent "real world" data, such as
pictures and sound, in digital format. To see how this might be possible, let us take a
(very) simple picture as presented in Fig. 2.1:

To convert this picture into digital data (*digitise* it) we first divide the picture into
numerous small picture elements (*pixels*) and then assign a binary value or sequence
of bits to each individual picture element. In the case of our simple drawing above, a
single bit per pixel will suffice (the value determined by whether or not some part of
the pixel contains some part of the line drawing); for a more sophisticated image,
such as for example a colour photograph, numerous bits will be needed per pixel.

For example; if each pixel could contain 1 of 256 different colours, 8 bits per
pixel would be needed, and so on. [In practice, generally fewer bits will actually be
required because various sophisticated compression algorithms are available which
we do not need to go into here.] Obviously, the more picture elements we use, the
more accurate our rendition of the original artwork.

So, for our masterpiece above the binary representation would be 00110000... if
we adopt the convention of starting our coding at the top left of the picture and using
1 for black, and 0 for white. Figure 2.2 above represents this digitised artwork (in slightly modified form). Assuming no compression our picture would
consume a total of 286 bits of our computer's memory, or 36 *bytes,* where 1 byte
is a chunk of 8 bits. Clearly the finer the "mesh" we use, the greater our accuracy of
representation of the pictorial content, and correspondingly the greater the memory
requirement.

We can approach the digitisation of more elaborate forms of real-world data, such as sound and moving pictures in a similar fashion; however, in these cases we also need to digitise the temporal component as well, dividing time into discrete tiny individual slices.

2.3   Binary Logic and Computation

2.3.1   Processing

Processing generally involves *change*. And so, processing in a binary digital computer, where data (of many different types) is represented in binary format, involves, at its base, the change of bit values, according to a set of instructions. In its most basic form, we may envisage a device with a single binary input, producing (after a short delay) a single binary output. How many different possible devices of this nature are there? Well, with a single binary input **A** there are four output combinations, as shown in Table 2.8.

Here we have labelled the four possible functions based on their outputs; two of these are trivial, producing either the output 0 or 1 irrespective of the value of the input **A**, another output simply reproduces the output as a copy of the input **A**. The final function, NOT **A** (third output column) has a more interesting function; it flips or *toggles* the input—so 0 becomes 1, and 1 becomes 0. We will see one of the potentially useful applications of this final function shortly.

For a binary device with two inputs (and a single output) the situation becomes a little more interesting and complicated. Here we have 16 potential devices, each with a different function. Some of these are elucidated in Table 2.9 below.

Here we have labelled 12 of the possible functions that can be generated by combinations of two binary inputs, **A** and **B**—the labelling of possible functions of the other four columns is left to the reader as an exercise. The methodology behind the names given to the other functions in the above table does require some explanation.

2.3.2   Arithmetic and Logical Operations

In a digital computer two core functionalities are needed: firstly, the ability to perform arithmetic calculations quickly and accurately, and secondly the ability to

Table 2.8 All possible binary single-input, single-output binary devices

Input	Outputs			
A	0	A	NOT A	1
0	0	0	1	1
1	0	1	0	1

Table 2.9 All possible two-input one-output binary devices

Inputs		Outputs															
A	B	0	AND		A		B	XOR	OR	NOR	A=B	NOT B		NOT A		NAND	1
0	0	0	0	0	0	0	0	0	0	1	1	1	1	1	1	1	1
0	1	0	0	0	0	1	1	1	1	0	0	0	0	1	1	1	1
1	0	0	0	1	1	0	0	1	1	0	0	1	1	0	0	1	1
1	1	0	1	0	1	0	1	0	1	0	1	0	1	0	1	0	1

perform logical operations. The latter is what gives the computer its decision-making capabilities.

Because of the fundamental nature of the processing carried out by some of the functions in Table 2.9, several of them are given their own unique graphical symbols, four of which are reproduced here in Fig. 2.3.

To illustrate important arithmetic capabilities in a binary computer, the addition operator is key. Binary multiplication is relatively straightforward; as there are only two numbers, 0 and 1, multiplication by 1 simply reproduces the number, and multiplication by 0 (of any number, in any number base) is 0, so multiplication in binary simply reduces to the addition of a number to itself a number of times, each time switched one digit to its left. We reproduce the binary addition and multiplication tables below.

It is clear from the above tables that the addition (sum) function is carried out by the function labelled XOR in Table 2.10. It is also clear that the carry function, as well as the multiplication function is implemented by the function labelled AND in this table.

Fig. 2.3 Symbols for some basic binary functions

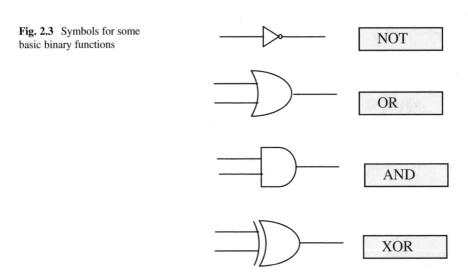

Table 2.10 Binary addition and multiplication tables

Addition	0	1
0	0	1
1	1	0 carry 1

Multiplication	0	1
0	0	0
1	0	1

Why label these functions AND, OR, XOR, etc.? Well, this derives from another interpretation that we can place on the two states represented in a modern binary digital computer. Normally we assign the values 0 and 1 to these states, but an alternative interpretation would be to interpret one state (1) as the logical value TRUE, and the other state (0) as FALSE. With this in mind, our labelling of the 16 possible functions that can be generated by the different combinations of two binary digits becomes a little clearer. For example, the AND function (which happens to correspond to the multiplication operator above) has as its output true in the case where A AND B are both true. The NAND function stands for NOT-AND; so the output is true (1) in all cases except where both inputs are true.

The XOR function stands for eXclusive-OR; this function has as its output true (1) in the case where either the A input OR the B input is true, but *excludes* the case where both are true.

2.3.3 Binary Addition

Addition is the fundamental arithmetic operation; with addition available to us multiplication and subtraction follow on shortly, as we will see soon. For multi-digit addition in any number base it is necessary to take account of not just the addition of two numbers in that base, but also any potential carry in value that might have been generated previously. For decimal addition this can get quite complicated; for binary addition the situation is a little more straightforward—it is represented in tabular form in Table 2.11.

This type of table—one which lists the outputs expected for particular combinations of the inputs—is sometimes called a *truth table*. The origins of this term are clear if you interpret the outputs from the table as the indicators of the truth or falsity of particular input propositions (where 1 = true, 0 = false). In order to perform multi-bit binary addition, we need to take account of not just the 2 bits to be added, but also a potential carry-in from a previous column. We reproduce in Table x.y the truth table for a device for the addition of two numbers A and B together with a carry-in value C_{in}. This is sometimes called a *full adder* (as opposed to a *half adder* for a device that omits this carry-in input).

Table 2.11 Full adder truth table

Row	A	B	C_{in}	Sum	C_{out}
1	0	0	0	0	0
2	0	0	1	1	0
3	0	1	0	1	0
4	0	1	1	0	1
5	1	0	0	1	0
6	1	0	1	0	1
7	1	1	0	0	1
8	1	1	1	1	1

To interpret the contents of this table let us examine one of the rows first. Looking at the sum term on row 2 it is clear that this value is true (=1) when the A input is false (=0), the B input is also false (=0), and the C_{in} input is true. However, the same is also true in rows 3, 5, and 8. Extrapolating from this we get:

Sum (is true when) =

(A is false AND B is false AND C_{in} is true)

OR

(A is false AND B is true AND C_{in} is false)

OR

(A is true AND B is false AND C_{in} is false)

OR

(A is true AND B is true AND C_{in} is true)

giving, in shorthand notation:

$$Sum = \overline{A}.\overline{B}.C_{in} + \overline{A}.B.\overline{C}_{in} + A.\overline{B}.\overline{C}_{in} + A.B.C_{in}$$

Here, the "+" symbol represents the OR function, the "." symbol represents AND, and a bar over the top of one of the operands represents the inverse of that value. The differentiation between the use of the + symbol used to represent the OR operator as opposed to addition is generally clear from the context; similarly for the "." operator in its use to represent either the AND operator or multiplication.

Similarly, we can generate the C_{out} result as

$$C_{out} = \overline{A}.B.C_{in} + A.\overline{B}.C_{in} + A.B.\overline{C}_{in} + A.B.C_{in}$$

Without going to the trouble of actually drawing out the circuit diagram it is clear that, in total, without any simplification we require a total of eight 3-input AND gates, and two 4-input OR gates, together with three inverters. This gives us a rough measure of the total complexity of the circuitry involved.

2.3.4 Karnaugh Maps

In order to interpret these results more fully a number of possible approaches can be used. In our analysis we use the Karnaugh-map (or *K-map*) approach. We will apply this approach in both the two-state (binary) situation, as used in almost all modern computer systems, and also shortly to the case of a hypothetical ternary computing system.

The K-map approach simply involves taking the data as contained in the truth table, and mapping this on a grid, where each square on the grid corresponds to a

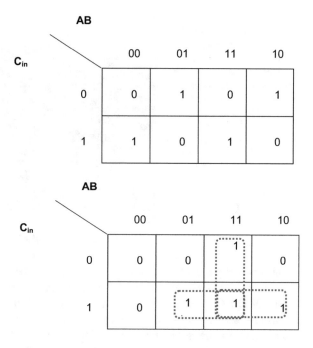

Fig. 2.4 Karnaugh maps generated for the sum (above) and carry-out (below) functions of the one-bit full-adder

particular combination of input values. So, for a three-input function, as in the full adder described above, the grid has 8 values; for a four-input function it would have 16 values, and so on. [K-maps are generally not applied to functions of more than six variables as it becomes quite difficult to visualise the results at this point].

Once the function is mapped, we look for groupings or blocks of 1's, of size 1, 2, 4, 8, 16, etc. These blocks may overlap. In general, we are looking for the smallest number of the largest-size blocks, such that all the 1's (and no 0's) are covered.

The K-maps shown in Fig. 2.4 are generated for the sum and the carry-out respectively for the full adder as described above. You may notice that in the labelling of the horizontal squares we do not progress in normal binary counting format (00,01,10,11), but rather (00,01,11,10). This is to create a situation where there is just a single-bit difference between adjacent columns. This numbering regime allows us to easily identify relevant regions in the graph—in the four-variable case a similar numbering system is used also for the horizontal columns. Also, although it does not arise in the cases we discuss here, the left and right sides of the map are deemed adjacent, as are the top and bottom edges.

From the K-map representation of the sum it is clear that no simplification is possible; i.e. it is not possible to form groupings of more than a single 1. However, in the case of the carry-out function we can clearly identify three groups. These groups happen to be overlapping, but this is no problem. The simplified function (from the K-map) for the carry-out is

Fig. 2.5 The carry-out function of a binary full adder implemented using simple switching circuitry. Here the bottom two switches are set in the down position, representing the value 1 (TRUE), the top switch is set in the up position, representing the value 0 (FALSE). The bulb at the top left is lit, correctly signifying an output of 1 (TRUE)

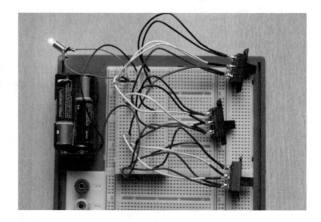

$$C_{out} = A.C_{in} + A.B. + B.C_{in}.$$

This is a significant reduction in the complexity of the required logic circuitry (from four 3-input AND gates and a 4-input OR gate, to three 2-input AND gates and a 3-input OR gate. It also gives us a better overall insight into the computations involved—we simply AND together each of the three combinations of the two inputs and pass the results through a 3-input OR gate. Figure 2.5 demonstrates the implementation of the carry function of a binary full adder using simple switching circuitry and using the simplified K-map-generated function.

2.3.5 Multi-bit Addition

It is now a straightforward procedure to perform multi-bit addition, simply by combining a number of one-bit adders in sequence, with the carry-out of one adder forming the carry-in input of the next.

The example in Fig. 2.6 shows a 3-bit adder circuit, this circuit adds two 3-bit numbers $A_2A_1A_0$ and $B_2B_1B_0$ to produce a result $S_2S_1S_0$; it is clear that this approach could be extended to 8, 16, 32, or as many bits as we wish. The carry-out generated by each full adder forms the carry-in input for the adder immediately to its left in the diagram. The rightmost adder will not receive any carry-in input, so its circuitry may be simplified by implementing this as a half adder. If the leftmost adder produces a carry output this cannot be handled by the system as it stands; if this occurs this bit is sometimes called an *overflow* (OV) bit.

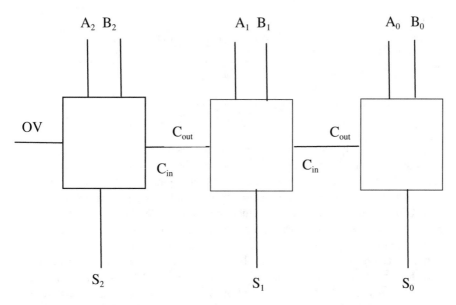

Fig. 2.6   A 3-bit full adder

Table 2.12 Odometer readings reading back from 0 miles (or kilometres)	
	00000
	00001
	00002
	. . . ..
	99998
	99999
	00000
	00001
	00002

2.3.6   Subtractors and Multipliers

The device we have just seen how to construct is an exceptionally useful one, especially given that the operations of subtraction and multiplication are very closely related to the addition operator. Division is a little more complicated; indeed, it has been argued that the complications involved in the implementation of the division operator in ternary logic are one of the major arguments against the adoption of ternary rather than binary logic. But this is a topic we will come to shortly!

For now, it is useful to see that binary subtraction can be performed using the simple equivalence

Table 2.13 Negative binary numbers

Bit pattern	000	001	010	011	100	101	110	111
Unsigned	0	1	2	3	4	5	6	7
Signed	0	1	2	3	−4	−3	−2	−1

$$A - B \equiv A + (-B)$$

So, in order to do subtraction all we have to do is to be able to create the negative equivalent of one of the variables. This is fine—but how do we now represent the negative of B, without introducing another separate symbol to represent the negative sign? We can arrive at an answer by looking at the ordinary car odometer. Let us start at 00000 (assuming that your odometer goes from 00000 to 99999 miles or kilometres; most go a little further than this...). So, we have as in Table 2.12.

And so on. Of course, the point here is that once we go beyond the range of our odometer (corresponding broadly to the computer's word size in our analogy) we reset to zero, and just start again. So, in a real sense 99999 is actually *less* than 00001, in fact 2 less, so we can view 99999 as −1. This is known as the *10's complement* representation of negative numbers. We can extend the same principle to binary numbers; here it is known as *2's complement*.

So, for 3-bit binary we have the signed negative numbers as illustrated in Table 2.13.

Of course, the available range (in this case eight distinct values) is now split in two, one section positive, and the other section negative values, and the question then arises where we draw the line between the (large) negative and positive numbers. By convention 0 is taken as a positive number, so, splitting the range evenly, we go from −4 to +3 for 3-bit binary, or from −128 to +127 for 8-bit, and so on.

There remains one final question: how can we quickly and easily go from the positive binary representation of a number to its negative equivalent? On consideration, we can observe that to go from +3 (011) to −3 (101) in the table above so we could simply invert all of the bits in the number, giving us 100 (−4), which is 1 less than the number we want; a quick check through the table above shows that this will work in every case. We can then update this to the correct number simply by adding 1. This now gives us a generalised method for the representation of negative numbers without the need for an extra symbol and allows for a simple and direct method of subtracting numbers.

Multiplication is the final arithmetic operator we will consider here. As we mentioned earlier, multiplication in binary is particularly straightforward, when we consider that multiplying a number by 1 simply involves reproducing that number; and multiplication by 0 of any number results in 0. Multiplication in binary then reduces to the addition of the multiplicand to itself a number of times; each time shifted a certain amount to the left, as determined by the multiplier below.

Fig. 2.7 A 2-4 decoder

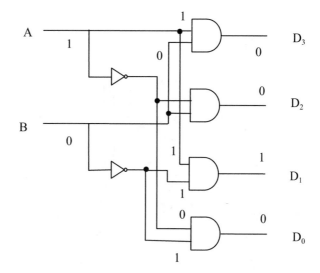

2.3.7   *Decoders and Multiplexers*

Let us take the time now to briefly describe two particularly interesting combinational circuits, the decoder and the multiplexer. A decoder takes an *n*-bit number as input and uses this to set equal to 1 exactly one of its 2^n outputs. This device can be particularly useful where we wish to have just a single component of a complex circuit operational at any one time.

In the decoder shown in Fig. 2.7, we take two inputs, A and B, and by generating the four possible combinations of these inputs we ensure that one, and only one, of the output lines will be set to 1. For example, if the input A is 1 (true) and the input B is 0 (false), then the output D_1 will be 1 (true) with all of the other outputs equal to 0. (Not all inputs to the OR gates are shown for reasons of clarity.)

Another very useful combinational device is the multiplexer. A multiplexer is essentially an electronic switch, and as, on one level, the core of a digital computer can be seen as a vast array of switches driving other switches, we can get a clear understanding of its utility in the construction of computational circuitry.

In the four-input multiplexer shown in Fig. 2.8, one of the four data inputs on the left is routed through to the output. The input selected is determined by the two control inputs C_0 and C_1. The operation of this device is clear if you consider the two control inputs and the leftmost column of AND gates to form an embedded decoder structure. So, one (and only one) of these four gates will have an output of 1. The second column of AND gates then takes each of these outputs in turn and combines them with one of the four data inputs. The end result is that only one of the second row of AND gates can have an output of 1, and whether it does or not is determined by its associated data input. This should be clear from the following brief discussion of *open* and *closed* gates.

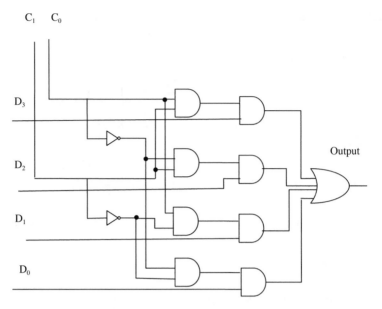

Fig. 2.8 A four-input multiplexer

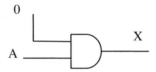

Fig. 2.9 A "closed" AND gate

2.3.8   Gates: Open and Closed

The AND gate can be looked at, in a sense, as a form of physical gate, but instead of allowing cars or sheep through the gate it allows through bits.

Consider the position shown in Fig. 2.9.

Given an input A, what is the output X?

After a moment's consideration we see that whatever A's input value, the output X is 0. In a certain sense here, we can say that the "gate" is closed.

Now consider the position in Fig. 2.10:

In this case what value do we place on the output X? Again, after a moment's reflection we see that, in this case, the X output should follow whatever value is presented at the input: $X = A$. The "gate" is now open, allowing bit values to flow through freely from input to output.

Getting back to our multiplexer, we can then finally combine the four outputs using a 4-input OR gate, as illustrated in Fig. 2.8. We could, in fact, replace the eight

Fig. 2.10 An "open"
AND gate

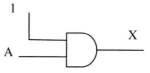

2-input AND gates with four 3-input gates, however the operation is, perhaps, a little clearer as illustrated in the diagram.

2.3.9 The Arithmetic Logic Unit

We will now see how to build an exceptionally useful device—an *Arithmetic Logic Unit* (ALU). While the combinational circuits discussed so far perform a single function, an ALU is a multi-functional device that can be told what function, either arithmetic or logical, we wish it to perform. In this sense it is on a basic level a *programmable* device and forms a core component of the computer's *Central Processing Unit* (CPU). In order to demonstrate the principle of operation of an ALU without getting bogged down in complex circuitry, we will show how to build a simple ALU with three logic functions, OR, XOR, and NOT, and just a single arithmetic operation: addition. The three logic functions are contained within the *Logical Unit* (LU), and the addition operator, implemented using the circuitry discussed earlier, forms the *Arithmetic Unit* (AU). All four functions are performed in parallel; the function finally forming the output of the device is then selected using a 4-input multiplexer as illustrated in Fig. 2.11.

So, depending on the values of F_0 and F_1, one of these four operations are selected, and the correct output is then generated. [In order to simplify the overall diagram, we do not represent the carry-in and carry-out lines associated with the arithmetic unit.] Figure 2.12 shows a simple 1-bit ALU constructed by the author using relay logic and capable of performing the four functions described above. The two binary inputs A and B are selected using the miniature lever switches in the bottom centre of the picture. The carry-in input is selected using the switch at the bottom left. The function (arithmetic or logical) to be performed is selected using the two small push-button switches at the bottom right of the photograph.

2.3.10 Multi-bit General Computation

Of course, the device outlined in the previous section, while undoubtedly multifunctional, only performs operations on a single pair of input values at a time. To build a truly useful device we must be able to perform arithmetic and logical operations on a number of bits at one time, typically 32 or 64 bits on a modern binary digital computer.

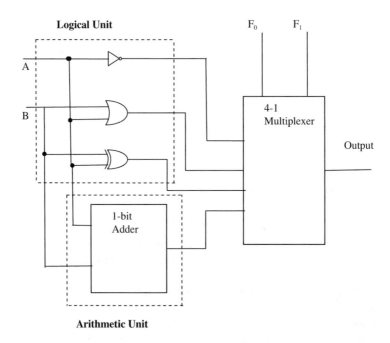

Fig. 2.11 A one-bit ALU implementing the functions of NOT, OR, XOR, and addition

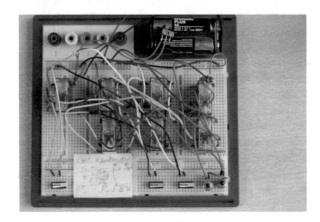

Fig. 2.12 A simple 1-bit ALU capable of performing the addition function (full adder) and the logical operations NOT, AND, and OR

In principle this can be easily achieved. If, instead of a single ALU as discussed in the previous section, we have 32 (or 64) such devices connected in such a fashion that the carry-out of one device forms the carry-in of the next device in line (as in the multi-bit adder), then we have, at least in principle, the capability to carry out true multi-bit computations.

2.3.11 With One of These I Can Get as Many of Those as I Want

The good news is that we do not, now, have to go on to look at devices of ever-increasing complexity. As the old joke goes: "with one of these I can get as many of those as I want". Of course, this does not preclude hardware developers from including additional, highly complex, digital circuitry to modern computers in order to increase speed and efficiency, especially as the rate of clock speed increases appears to be plateauing. However, the core ability of the computer's CPU is still to be able to perform any arithmetic or logical function it is required to do. The ALU is a real "programmable device" in the sense that you tell it what function it is to perform and provide the inputs, and a very short while later the correct result is generated.

2.4 Memory

As stated earlier, along with the ability to process data effectively, the second core element of a computer system is the ability to store intermediate and final results generated in some form of computer memory.

An example of a simple memory element is given in Fig. 2.13, as implemented using two NOR gates, connected together in a rather unusual way. They are connected in such a fashion that one of the inputs to each device is provided by the output from the other device. This configuration is characteristic of feedback control devices in general, where some element of the output of the device forms part of the input also.

Let's see what happens here—assume S and R are both 0 and also assume that $Q = 0$. This is a stable state for this device. Now assume S, R are both 0 as before—this time assume $Q = 1$. Again, this is a stable state. Any state with both outputs the same is inconsistent.

However, no matter which of these two states we're in, if we apply a 1 to the S (*Set*) input then Q becomes 1; setting R (*Reset*) to 1 forces the output to 0. The key issue here is that the circuit *remembers* whether S or R was last on, and thus can form the basis for a simple 1-bit memory element.

Fig. 2.13 The SR latch

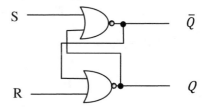

2.5 Synchronous Versus Asynchronous Computation

We have now seen how to construct combinational logic units, able to generate arbitrary functions based on their binary input values. We have also seen how to construct a device—the ALU—which is able to perform any of a range of functions at will, without specific rewiring. However, in order to have true computational capabilities, in line with our earlier discussions, it must be possible to perform a *sequence* of operations, one after the other, with each operation following on from the previous one. There are two broad methods of implementing this *sequential logic*—*synchronous*, and *asynchronous*.

Synchronous logic circuitry, as used in the vast majority of modern computing devices; involves the use of a separate *clock* mechanism, which specifies the duration of each individual state, and marks the transition from one state to another. In the case of an asynchronous unit the transitions between one state and another in the logic circuitry are essentially determined by the delays between the logic components themselves, as determined by the particular technology used to implement the logic—be it electronic, as used in the vast majority of computers, or mechanical, or electromechanical (using relays, etc.), or even, for example, fluidic components.

2.6 Reasons for the Power of Modern Computers

There are two main factors that contribute to the vast power of computing devices today, as opposed to, for example, 20 years ago. Firstly, the hugely improved ability to cram many miniature switching devices into increasingly tiny spaces. The current state of the art is 20 nm. This means that the space required to implement a single gate (AND, OR, etc.,) is on the order of 20 *billionths* of a metre square!

The second factor contributing to this huge increase in computing power is the great leap forward over these same years in the speed of operation of the basic core computing devices, as measured by the clock speed of the computer. Current computer clock speeds range in the 2–4 GHz range; that is the number of beats of the clock is in the region of 2–4 *billion* cycles per second. By contrast, the basic neuronal cycle of the human brain is on the order of milliseconds (thousandths of a second). We can clearly see from these figures some of the factors that afford modern computers their vast computational ability, within a relatively small form factor (space).

Figure 2.14 shows the timing diagram for a clock operating at 1 Hz—that is, one cycle per second. In order for a synchronous circuit to operate it is necessary to store

Fig. 2.14 Clock timing diagram

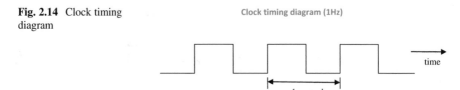

Clock timing diagram (1Hz)

the results generated by the combinational circuitry in some form of computer *memory*, in order to provide the next inputs in order to continue the calculations. In asynchronous circuitry there is no such requirement as the memory facility is provided inherently by the internal logic delays within the circuitry. (Of course, this is not to say that memory is not required to store instructions and data within asynchronous machines.)

2.7 Yes, No, Maybe?

Now, at this point it is probably helpful to point out that the binary perspective on the world, while undoubtedly very useful, is not the only possible one; or indeed the ideal. For example, if asked whether Prague was the capital of the Czech Republic, I could either answer "Yes" (1), it is (which I happen to know to be true, as I have visited this very interesting and historical city), or "No" (0), which is incorrect; or I could *also* answer "I don't actually know" (D), which might be the case for many people, certainly those outside the greater European area.

The point here is that the pure binary response (true/false) does not allow for this extra graduation, which ternary logic (base 3) does explicitly allow for. (Of course, there is also the whole research area of Fuzzy Logic, which falls outside the general remit of this text.)

2.8 Ternary Computation

> Popular music is usually based on the number 2 and then filled in with fabrics, colors, effects and technical wizardry to make a point. But the total effect is usually depressing and oppressive and a dead end which at most can only last in a nostalgic way.... I don't know my why the number 3 is more metaphysically powerful than the number 2, but it is.
> —Bob Dylan (2004) *Chronicles*

We will now take a brief excursion into the exotic world of ternary computation. Our reason for this excursion is twofold; firstly, to break away from the notion that digital computers have to be binary by nature—they do not—although the vast majority of modern computers operate on binary principles. Secondly, the use of ternary, particularly balanced ternary, comes with several advantages, and has a particular elegance not possessed by binary. The well-known and highly respected computer scientist Donald Knuth, recipient of the 1974 Turing Award (the "Nobel Prize of computing") once described the balanced ternary notation as "Probably the prettiest number system of all" (Knuth 1981).

We will come to balanced ternary shortly; however, we look initially at standard ternary notation, that is using the digits [0, 1, 2], as opposed to the binary [0, 1]. What advantage might this system have over standard binary notation? Well, as we saw earlier, in looking at the decimal number 645_{10} in different number bases, we will always require fewer trits than bits to represent a number—six (212220_3), as opposed to ten (10100001011_2), in this case. But, of course, this is not the whole story.

Duodecimal (base 12) or hexadecimal (base 16) require even fewer digits; generally, fewer even than the decimal equivalent of the number. In fact, clearly, the higher the radix, the fewer the number of digits required to represent numbers or other data. But, of course, the use of these higher radices comes at a cost—the cost being the extra complexity required in the computing machinery in order to represent/store/process each digit. So, a true measure of the cost of representing data in our computer must be some *combination* of the *number* of digits required to store this data together with the *cost* of representing each of these individual digits.

A straightforward measure here would be to minimize the cost of the product of these two quantities $C = r*n$, where C is the overall cost, r is the radix, and n is the number of digits employed. A simple example might help here to illustrate the advantage of the ternary system given the (very plausible) cost/complexity measure proposed.

2.8.1   An Example of the Ternary "Advantage"

Let us assume $C = 6$. Then, given integer values of r and n we can look at the different possibilities, taking for convenience just those radices that divide the number 6 evenly. From the unary case (which is not really a positional notation), which allows for six possible combinations (1, 11, 111, 1111, 11111, 111111), to the senary (base 6) case, again allowing for six combinations (0, 1, 2, 3, 4, 5), there are four possible number bases in total. These are (in order of radix): unary, binary, ternary, and senary. We have just discussed the unary and the senary cases. Taking binary, $r = 2$ so $n = 3$, and we have three binary digits to play with, giving a total of 2^3 possible combinations, or 8. For ternary the equivalent figure is 3^2, or 9, an improvement on the binary figure (and, of course, also on the unary and the senary case).

2.8.2   Another Example...

But we might say, this is just artefactual; we picked a cost/complexity factor of 6, perhaps this was specifically chosen in order to show off the advantage of ternary.

Let's take another value: $C = 30$. Here (again just taking radices that are divisors of C), we have $r = 1, 2, 3, 5, 6, 10, 15, 30$. Again, ternary comes out on top; for example we have both unary and base 30 with 30 values in total, decimal with 1000 (10^3), binary with 32,768 (2^{15}), and ternary again coming out on top with 59,049 (3^{10}).

So, based on this evaluation of the cost/complexity of the representation of numerical (or other) data, ternary appears to come out on top in most cases. In fact, in virtually all cases ternary is superior, it is only for low values of the C function that either binary or quaternary (base 4) is better. These cases do not, of course, correspond to the representation of real-life data where normally we wish to represent and to process multi-digit data. In fact, when one does a theoretical analysis of the results it turns out that the transcendental number e provides the optimum number base; unfortunately e (2.718281828) is not an integer, and it turns out that the closest number base to e happens to be 3 (ternary). The interested reader should see Appendix A "What is the most efficient number base" for a more detailed discussion of this topic, where we also look at the implications of different choices of cost/complexity functions.

2.8.3   Balanced Ternary

The balanced ternary system, as referred to by Knuth above, involves replacing the three digits 0, 1, 2 in standard ternary by the numbers -1, 0, $+1$, sometimes written as $\bar{1}$, 0, 1, or simply $-$, 0, $+$. In terms of representing numbers this system operates in an identical fashion to standard positional notation, except for the fact that in some cases we subtract rather than add values to obtain the overall numerical value being represented. As an illustration, here are the numbers -4_{10} to $+4_{10}$ as represented in balanced ternary notation in Table 2.14.

So, for example, the number 2, represented by $1\bar{1}$, is decoded as $(1 \times 3^1 - 1 \times 3^0)$ $= (3-1) = 2$.

2.8.4   Logical Operations

We can also construct logical operations using ternary. Table 2.15 gives is the truth table for the logical OR operation in ternary using the more conventional -1, 0, $+1$ symbols.

Table 2.14 Balanced ternary representation of the integers -4 to $+4$

Decimal value	Balanced ternary
-4	$\bar{1}\,\bar{1}$
-3	$\bar{1}\,0$
-2	$\bar{1}\,1$
-1	$0\,\bar{1}$
0	$0\,0$
1	$0\,1$
2	$1\,\bar{1}$
3	$1\,0$
4	$1\,1$

Table 2.15 Ternary logical OR function

A/B	-1	0	$+1$
-1	-1	0	$+1$
0	0	0	$+1$
$+1$	$+1$	$+1$	$+1$

We can interpret the values in this table in the following fashion. We assign the value TRUE (as in binary) to +1, the value FALSE to −1, and the value UNKNOWN (or uncertain) to the digit 0. With these interpretations in place the truth table makes perfect sense. For example, if both inputs to the ternary device have the value −1 (FALSE), then the output is FALSE, as we would expect, and as is the case for the binary OR gate. If, on the other hand, one of the inputs is TRUE, and the other UNKNOWN, the output will have the value +1 (TRUE). On a moment's reflection this is correct—while the UNKNOWN input may have the actual value TRUE or FALSE, this is immaterial as we know that the other input is true, and the output from our OR gate should be true irrespective of the first value. However, if one input is FALSE and the other UNKNOWN, then the output in this case is UNKNOWN, as the final result will, in this case, be dependent on the UNKNOWN input. Of course, in the case where both inputs are TRUE, the output is also TRUE.

2.9   Design and Construction of a Balanced Ternary ALU

I will now outline the details of the construction of a ternary arithmetic logic device (ALU), which the reader with a little time on their hands and some dedication may well wish to try their hands to construct. All of the components used are readily available from most electronic component retailers, and the device, once constructed, should serve as an excellent educational tool demonstrating basic principles of ternary arithmetic and logic. I have used this ALU myself to good effect in illustrating the basic principles of non-binary arithmetic and logic operations to first-year undergraduate computer science students. Initially I was reticent about its use in this regard—these students had only recently been introduced to binary systems and I did not wish to cause confusion; however, far from this being the case I think that following the demonstration many students, especially the more gifted, found a deeper understanding and much greater interest in the whole topic of digital computation.

2.9.1   Details of the Ternary ALU Construction

Here I describe in more detail how to construct a balanced ternary ALU, based on the principles outlined earlier, using relay logic. (It should be noted that this section may be omitted on a first reading of this text, as little of what follows is required reading for comprehension of the following chapters.)

Of course, the problem with the use of present-day relays in the implementation of ternary switching circuitry is that the vast majority of relays nowadays are essentially two-state in nature. A voltage of sufficient strength is applied to an actuating coil, which magnetizes a metal core at its interior, which then draws a

Fig. 2.15 The completed ternary ALU. This device computes the addition function for two ternary values A and B, represented in balanced ternary form in conjunction with a carry-in value C_{in}, also represented in balanced ternary. It also computes the ternary logical OR and NOT functions. It is currently shown computing the addition function based on input values $A = +1, B = 0$ and $C_{in} = +1$. This generates the output $+1, -1$, as represented by the green light on the left (representing $+1$), and the red light on the right (representing -1). The function select knob on the left is set to perform the addition function; the remaining three knobs (from left to right) comprise the A, B, and C_{in} inputs respectively. Each knob can be set in one of three positions, representing either a single trit, or in the case of the leftmost input, the function to be computed by the ALU. The device comprises a total of 19 DPDT relays, with associated wiring, and is assembled on three individual breadboards

miniature metallic lever towards it, thus making (or breaking) a connection. However, it is also possible that the moving lever/armature could be composed of a permanent magnet, so that the direction of movement of the armature would be dependent on the direction of current flow, thus allowing the implementation of simple ternary logic gates.

While not readily available nowadays the author did manage, as a point of principle, to successfully construct a device sensitive to current direction by the judicious insertion of a permanent magnet, as described, into the innards of a standard relay mechanism. However, in practice, and for ease of implementation, standard relays were used in the final circuitry, in conjunction with diodes, which restricted the switching of the relay to current coming from just one particular direction. We outline the operation of the ALU below; for reasons of space, this discussion is quite restricted; the interested reader is referred to the original publication (Eaton 2012) for a detailed description. This will hopefully allow for the construction of your own version of such a device with a little study and application!

As illustrated in Fig. 2.15, this simple ternary ALU performs the arithmetic function of addition, together with the three logic functions, AND, OR, and NOT. It should be noted that this device is, of course, not the only device that is capable of implementing the ternary arithmetic and logical functions as discussed above. It is, however, to the author's knowledge, one of only a few, if not the only device in existence today to implement these functions in the manner described above in a physically realised device.

As mentioned above we direct the interested reader to the detailed discussion of this device in (Eaton 2012). However it will be helpful to outline the overall operation of this device, which we will do now.

2.9.2   Ternary Processing

Of the 27 possible unary (or *monadic*) functions of a single-input single-output ternary device we select 12 functions that are easily implementable using the diode relay logic outlined above, and we label these the *primary monadic functions*. We also introduce a notation where we label the conditions of a variable A being true, unknown, or false respectively by the following convenient symbolic notation: $\overrightarrow{A}\,\underline{A}$ $\overleftarrow{A}$. Using this notation, we outline the operation of these 12 monadic ternary functions in Table 2.16. For the ternary inverse function true becomes false, false becomes true, and the unknown value remains unchanged.

We represent the (ternary) inverse of these functions by the use of square brackets [ ], so $A = (+1, 0, -1)$ and $[A] = (-1, 0, +1)$.

As well as these 12 functions we also use two further dyadic (two-input) functions, which we term the *Functional-AND (F-AND)* and the *Functional-OR (F-OR)*, both of which are easily implementable in our system. The F-AND sets its output as true, only if both of its inputs are true. The F-OR function corresponds broadly to the OR function in binary logic. In this device, if either or both inputs are true with no false input the output is true. Conversely, if either or both inputs are false with no true input the output is false. Otherwise the output is 'unknown' (0). As in binary these functions are represented in shorthand by a dot (.) and a + sign respectively. Further details of the operation of these two functions are given in Eaton (2012). It can be demonstrated that all dyadic and higher-order ternary functions can be implemented by these two dyadic functions, in combination with the 12 primary monadic functions.

As an example, take a simple ternary inverter. We wish the output of the inverter to be true if the input is false, false if the input is true, and for no change if the input is in the unknown state. The corresponding equation is $\left[\overrightarrow{A}\right] + \overleftarrow{A}$, where the + symbol

Table 2.16 Primary monadic ternary functions.

A	$|A|$	$\underline{A}$	$\overrightarrow{A}$	$\overleftarrow{A}$	$\overrightarrow{\underline{A}}$	$\overleftarrow{\underline{A}}$	$[|A|]$	$[\underline{A}]$	$\left[\overrightarrow{A}\right]$	$\left[\overleftarrow{A}\right]$	$\left[\overrightarrow{\underline{A}}\right]$	$\left[\overleftarrow{\underline{A}}\right]$
+1	+1	0	+1	0	+1	0	−1	0	−1	0	−1	0
0	0	+1	0	0	+1	+1	0	−1	0	0	−1	−1
−1	+1	0	0	+1	0	+1	−1	0	0	−1	0	−1

refers to the aforementioned Functional-OR function. We can verify this by checking the rows in the table above for the constituent components of this equation.

2.9.3   K-Map Simplification of Balanced Ternary Functions

We can also apply a modified version of the Karnaugh mapping technique to the simplification of ternary logic. We can do this by the identification of the primary monadic functions on either a row or column basis from the truth table of the function to be simplified. To briefly illustrate this technique, let us take the standard ternary logical OR function discussed earlier, as shown in Table 2.17.

We are now looking for horizontal or vertical groupings of +1's or −1's such that all of the non-zero elements are covered. Here we have clearly identified three groupings—the first two are the horizontal red box corresponding to the region where the A input is true and the vertical red region corresponding to the region where the B input is true; both of these inputs result in an output of true. (As in standard K-maps, overlapping of regions is permitted; however, region sizes do not have to be powers of 2). The final grouping, outlined in blue in the upper left of the table, corresponds to the case where both the A and the B inputs are false; this results in an output of false. The output from the device in all other cases is 'unknown' (0). This simply translates to the function $\vec{A} + \vec{B} + \left[\overleftarrow{A}.\overleftarrow{B}\right]$.

2.9.4   Balanced Ternary Addition

We now turn our attention to the addition function, the core, as we have seen in the binary case, of all of the arithmetic functions apart from division. The situation is considerably more complicated in the ternary case, as, for a full ternary adder (an adder that adds together two ternary inputs, together with another possible carry-in input) we have a total of 27 combinations (as opposed to 8 in the binary case) to deal with. We enumerate these possibilities in tabular form in Table 2.18.

Table 2.17   K-map simplification of ternary logical OR function

A/B	−1	0	+1
−1	−1	0	+1
0	0	0	+1
+1	+1	+1	+1

Table 2.18 Balanced ternary full adder truth table

A	B	C_{in}	Sum	C_{out}
−1	−1	−1	0	−1
−1	−1	0	+1	−1
−1	−1	+1	−1	0
−1	0	−1	+1	−1
−1	0	0	−1	0
−1	0	+1	0	0
−1	+1	−1	−1	0
−1	+1	0	0	0
−1	+1	+1	+1	0
0	−1	−1	+1	−1
0	−1	0	−1	0
0	−1	+1	0	0
0	0	−1	−1	0
0	0	0	0	0
0	0	+1	+1	0
0	+1	−1	0	0
0	+1	0	+1	0
0	+1	+1	0	+1
+1	−1	−1	−1	0
+1	−1	0	0	0
+1	−1	+1	+1	0
+1	0	−1	0	0
+1	0	0	+1	0
+1	0	+1	−1	+1
+1	+1	−1	+1	0
+1	+1	0	−1	+1
+1	+1	+1	0	+1

And transferred over to modified K-map format we obtain the Karnaugh maps generated, as shown in Table 2.19.

Again, we look to identify the regions. Taking the sum, where the carry-in is 0 we generate the function

$$S_0 = \overleftarrow{A}.\overleftarrow{B} + \underline{A}.\overrightarrow{B} + \overrightarrow{A}.\underline{B} + \left[\overleftarrow{A}.\underline{B}\right] + \left[\underline{A}.\overleftarrow{B}\right] + \left[\overrightarrow{A}.\overrightarrow{B}\right]$$

The six relevant regions are circled in the left-hand middle diagram in the table above. In order to generate the sum output in the other two cases (carry-in $= -1$ and carry-in $= +1$) we note the similarity between these three tables; in fact the sum output generated when the carry-in is -1 is just the same as in the case where the carry-in is 0 except we 'rotate' the trits in the table to the left, so 0 becomes -1, -1 becomes $+1$, and $+1$ becomes 0. If we do the same thing, except rotating to the right

Table 2.19 Karnaugh maps generated for the ternary full adder function

Sum   **Carry-out**

Carry-in = -1

A/B	−1	0	+1
−1	0	+1	−1
0	+1	−1	0
+1	−1	0	+1

A/B	−1	0	+1
−1	−1	−1	0
0	−1	0	0
+1	0	0	0

Carry-in = 0

A/B	−1	0	+1
−1	+1	−1	0
0	-1	0	+1
+1	0	+1	−1

A/B	−1	0	+1
−1	−1	0	0
0	0	0	0
+1	0	0	+1

Carry-in=1

A/B	−1	0	+1
−1	−1	0	+1
0	0	+1	−1
+1	+1	−1	0

A/B	−1	0	+1
−1	0	0	0
0	0	0	+1
+1	0	+1	+1

instead, we generate the correct output in the case where the carry-in $= 1$. So, we have

$$Sum(Carryin = -1) = \overleftarrow{S_0} + \left[\underline{S_0} \right]$$

$$Sum(Carryin = 1) = \left[\overrightarrow{S_0} \right] + \underline{S_0}$$

In a similar vein we can generate the carry-out outputs; in the case of both the carry-in $= -1$ and carry-in $= +1$ we can use our modified K-map technique to good effect to generate simplified equations—the regions involved are circled in the tables above. The final equations generated are as below

$$CarryOut(Carryin = 1) = \overrightarrow{A}.\overrightarrow{B} + \overrightarrow{B}.\overrightarrow{A}$$

$$CarryOut(Carryin = -1) = \left[\overleftarrow{A}.\overleftarrow{B}\right] + \left[\overleftarrow{B}.\overleftarrow{A}\right]$$

$$CarryOut(Carryin = 0) = \overrightarrow{A}.\overrightarrow{B} + \left[\overleftarrow{A}.\overleftarrow{B}\right]$$

2.9.4.1   Input to, and Output from, the Ternary ALU

Input to the ternary ALU is via four 3-position rotary switches (See Fig. 2.15), three of which input the A, B, and optionally C_{in} (if the function to be performed is addition) values. The final switch sets the function; ternary full addition, ternary logical OR, or ternary NOT. The output from the ALU is via two bi-colour light emitting diodes (LEDs). Green represents a +1 (TRUE) value, red represents a −1 (FALSE) value, and no light represents 0 (don't know). The entire logic circuitry is implemented using 19 double-pole double-throw (DPDT) relays, which are transparent, allowing the switching action to be clearly observed, together with a selection of associated diodes.

2.9.5   Some Practical Details

It might be argued that ternary machinery is inherently more difficult to construct than binary. One could argue that while the logic circuitry may be more complex, the construction of a ternary ALU such as the one described here is not so complex as to be beyond the capability of the average electronics enthusiast. I would estimate the total cost of construction to be approximately 250 euros, the major portion of this cost being taken up by the relays themselves. Of course, the thrifty experimenter may be able to source cheaper (or perhaps second-hand) relays, thus considerably reducing their overall expenditure. Also, the cost of the device was increased because of the lack of availability of relays able to switch efficiently between three states, thus necessitating the use of two 2-state relays (plus associated diodes) in the place of a single 3-state relay. So, one would expect a significant reduction in costs if such 3-state relays were to be made readily available.

This ALU has proved itself robust in operation over several years, requiring minimal attention and kept simply covered with a light dust cover. It has been used to good effect in demonstrations over these years, mainly as a demonstration tool to first-year Computer Organisation students to illustrate clearly that effective computation is possible in other than a binary substrate—that indeed there may be more efficient and elegant mechanisms, as evinced by balanced ternary.

And this, indeed, gentle reader, is part of the reason for the inclusion of this discussion in the current text. The title of this book is *Computers, People, and*

Thought. Straight away most of us would consider computation to be constrained within the rigid narrow confines of modern binary digital computers. As, I hope, we have now seen, digital computation does not need to conform to these rigid boundaries.

However, there *are* boundaries with digital synchronous computing machinery; be it binary or ternary. As we have seen, at the core of digital computation is one simple idea or mechanic—*switches driving switches*.

2.10   Bringing It All Together

We now have the three core components of a synchronous digital computing system: combinatorial devices to perform (virtually) instantaneous calculations, a memory to store the results of those calculations, together with a sequence of instructions to be executed, the *program*, and a clock to step us in sequence through a sequence of computations as specified by the program (where this sequence may itself be stored in memory), and where intermediate values are also stored in memory. Now all we need is a mechanism that allows us to tie these components together—and the Algorithmic State Machine (ASM) charting methodology will allow us to do just that.

2.10.1   *Introduction to the Algorithmic State Machine Charting Methodology*

The Algorithmic State Machine (ASM) charting methodology is a powerful design technique for the design of sequential circuitry. The ASM methodology is normally applied to the design of binary state machines for implementation using binary *flip-flops* (simple memory elements based on the SR latch discussed earlier which also allow for a clock input) and associated circuitry, or some other hardware implementation. In our brief discussion we will extend the use of the methodology to the design of ternary state machines, for subsequent implementation using *flip-flap-flops* (the ternary equivalent of binary flip-flops) and associated ternary logic circuitry. We will apply the methodology initially to the design of a 4-state binary sequential logic circuit, and then to a nine-state ternary state machine, both with and without input. We will also illustrate the implementation of these devices using flip-flop and flip-flap-flop technologies. Discussion of the binary case follows here; the interested reader is then referred to Appendix B for an introduction to ternary state machine design.

2.10.2   The ASM Chart Description of Sequential Logic

In this section we will briefly describe the two core elements used in the ASM charting methodology (actually there are three in total, but for our overview presentation the three described are sufficient). These are the *state box* and the *decision box* as shown diagrammatically in Figs. 2.16 and 2.17.

2.10.3   Binary State Machine Design: A Machine with Input

We will now illustrate the implementation of a simple 3-state binary device with two inputs. The ASM chart for the device is given in Fig. 2.18. In order to implement this device we will require some underlying memory devices. We will use the JK flip-flop memory device in this instance. This device is broadly based on the SR latch discussed earlier, and has been used in the past in the implementation of many synchronous logic circuitry because of the relative simplicity of the circuitry generated. The overall operation of this device is defined in the *state table* in Table 2.20.

This table gives the next state of the memory device based on the four possible J and K inputs. From this table we can now generate the excitation table for this flip-flop as shown in Table 2.21. This table is useful because it tells us which J and K inputs are required in order to in order to effect changes from our *present state* (P) to the required *next state* (N).

Following this we generate the state assignment table, and the JK input maps as shown in Tables 2.22 and 2.23 respectively. In this instance, because we have two

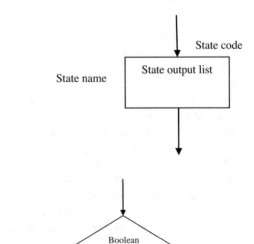

Fig. 2.16 The ASM State box

Fig. 2.17 The ASM decision box

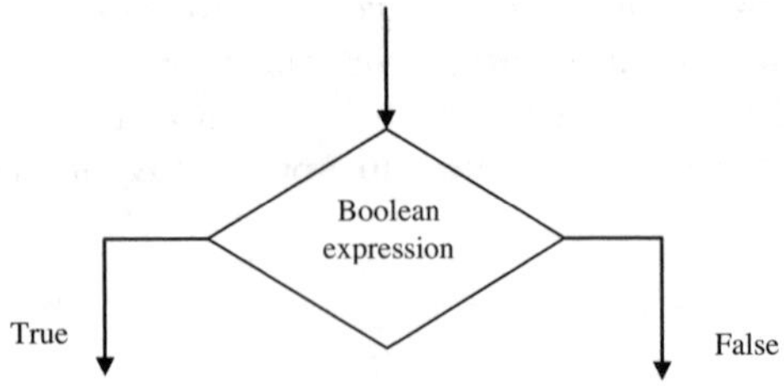

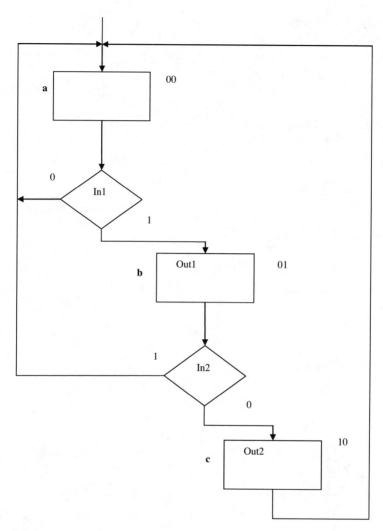

Fig. 2.18 ASM chart for a 3-state binary state machine with two inputs

Table 2.20 State table for JK flip-flop

J	K	Next state
0	0	No change
0	1	0
1	0	1
1	1	Toggle

inputs to our device, we use a technique called the *variable-entered map* (VEM) method. This simply means that we use the two input values as variables that are entered directly into the maps generated. From these maps we can then generate the final equations describing the sequential logic circuit generated.

Table 2.21 JK flip-flop excitation table

Present state (P)	Next state (N)	J	K
0	0	0	–
0	1	1	–
1	0	–	1
1	1	–	0

Table 2.22 State Assignment for machine with input

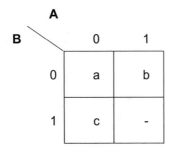

Table 2.23 JK input maps for machine with input

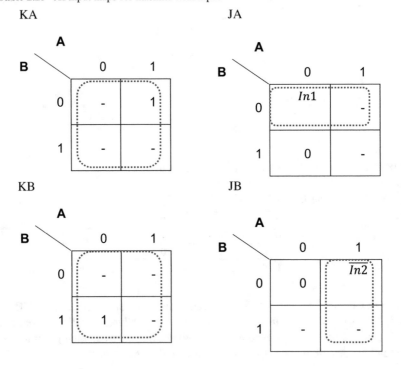

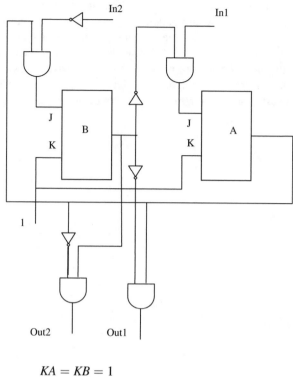

Fig. 2.19 JK flip-flop implementation of the 3-state machine with two inputs

$$KA = KB = 1$$

$$JA = \bar{B}.In1$$

$$JB = A.\overline{In2}$$

The corresponding circuit diagram is given in Fig. 2.19. Each flip-flop will also have a clock input, which is not shown for reasons of clarity.

2.11 Moving on

Just as the ALU is, in a sense, a programmable combinational logic circuit allowing us to implement a wide variety of logic functions without a change to the underlying hardware, so also we can construct a programmable device allowing us to implement a wide variety of state machines (possibly specified in ASM chart form). And, at its core a modern digital computer is a vastly complicated programmable finite state machine. Then, following on from this we then have the low-level languages, and also the high level languages that allow us to specify precisely, in an English-like language format exactly what operations we want the computer to perform.

And now we come to the end of this much condensed discussion of the fundamentals of digital computation— alas, space now precludes us from further detailed
discussion of this fascinating topic. Many books have been written on the topic to
which the interested reader may refer to expand their knowledge. However, hopefully now the witting reader should now have a basic understanding of the core
mechanisms of operation of a digital computer to help fuel, and also inspire, their
investigations of the further chapters of this book.

Chapter 3
People

The human body is placed, on the scale of magnitudes, halfway between the atom and the star. According to the size of the objects selected it appears either large or small... In reality, our spatial greatness or smallness is without importance. For what is specific of man has no physical dimensions.

—Alexis Carrel (1935). Man, the Unknown. *New York: Harper & Brothers.*

Life would be tragic—if it weren't funny.
—Stephen Hawking (1942–2018)

3.1 Where Do We Come from: Human Evolution

When God created mankind, he made them in the likeness of God. He created them male and female and blessed them. And he named them "Mankind" when they were created. When Adam had lived 130 years, he had a son in his own likeness, in his own image; and he named him Seth. After Seth was born, Adam lived 800 years and had other sons and daughters... When Seth had lived 105 years, he became the father of Enosh. After he became the father of Enosh, Seth lived 807 years and had other sons and daughters..... When Enosh had lived 90 years, he became the father of Kenan. After he became the father of Kenan, Enosh lived 815 years and had other sons and daughters...
  —The Book of Genesis

...hominids, along with their sole surviving representative, appear to have emerged not through a genetic revolution, but through a peaceful transition involving fairly large founding populations.
  —Klein, J., & Takahata, N. (2002). *Where do we come from? The molecular evidence for human descent.* Springer Science & Business Media.

It is now fairly generally accepted that the last common ancestor of humans and chimpanzees (our closest living animal relatives) lived approximately 5 million years ago. Since then many different species of *hominins* (members of the hominin tribe, which includes humans, and all of our now extinct ancestors back to the divergence from the last common ancestor of chimps, and also of bonobos) emerged,

© Springer Nature Switzerland AG 2020
M. Eaton, *Computers, People, and Thought*,
https://doi.org/10.1007/978-3-030-55300-5_3

only one of which is living in the present day, ourselves, *Homo sapiens* (from the Latin "wise man").

We should note here in passing, to avoid confusion, that the term *hominid* (as in the quote from Klein and Takahata above) used to have the same meaning as the current term hominin, however its meaning has now been broadened to include gorillas, orang-utans, chimpanzees, and bonobos.

What we call *modern* humans are generally accepted to have developed sometime between 200,000 and 50,000 years ago (200–50 ka). Modernity in this context can be defined in both biological and cultural terms. When evaluating biological modernity, we generally look to the fossil record and to hominins that were similar to us in terms of their brain size and their skeletons. Cultural modernity is more difficult to assess and derives from the archaeological record providing evidence of artistic and intellectual capabilities (Ackermann et al. 2016).

There is a broad general acceptance among many anthropologists of the so-called *Out of Africa* concept (also sometimes referred to as the *Recent African Origins* (RAO) model), in that in or around 200 ka in Africa there existed the ancestors of modern humans, who subsequently moved out into Europe and Asia replacing their nonmodern cousins (such as the Neanderthals) in the process. However, some would argue that there was no one defining point or 'short' interval in which modern humanness arose, but that this was rather a gradual process stretching over several hundred thousand years (McBrearty and Brooks 2000).

Others would argue that in or around 50–40 ka, a neural change, undetectable by archaeological means spurred the development of modern human behaviour, due to an alteration in brain organisation, rather than brain size. What is indisputable is that, for whatever reason, it is only after about 50 ka that the behavioural markers of modern humans become relatively commonplace in nature (Klein 2013).

Another hypothesis is the so-called *Multiregional Evolution* (MRE) model which holds that Homo Sapiens began to evolve around 2 million years ago and involved an interbreeding network of this species over the intervening period (Galway-Witham and Stringer 2018). Recent evidence confirms interbreeding (also known as hybridization) between H. Sapiens and other hominin species including Neanderthals, which lends support to the claim by some anthropologists that H. Sapiens and Neanderthals were, in fact, members of the same species. This is supported by the recent recognition of the Neanderthal origins of cave paintings in Spain (Hoffmann et al. 2018).

Current thinking among many researchers is that an intermediate theory embracing aspects of both the RAO and MRE models is most probably correct, with the *assimilation model* (AM) and the *RAO model with hybridisation* (RAOWH) emerging as likely contenders (Galway-Witham and Stringer 2018).

While this is certainly a fascinating topic and currently the subject of much hot debate, space precludes us from entering into further details here, the interested reader is encouraged to consult further the references cited. What is *not* in debate is that today humans occupy a unique role in the ecology of our planet with their ability to alter dramatically their living environments with the ability to build towns and vast cities, to alter drastically the courses of waterways, and to cause lasting changes to their environment. Humans also have, uniquely, now the ability to alter directly, through genetic engineering and without waiting for the evolutionary process, the ability to radically alter the future shape of their species.

So while the evolution of humans to the present day has, in major part, remained out of human direct control for vast millennia in time up to the relatively recent past, the future of humankind not just in the next decades and centuries but also into future millennia, falls very much within human hands and the decisions that will be made in the coming years.

3.2   The Transhumanist Agenda

Transhumanism involves the melding of technology and human biology into a form that its proponents say will transcend humanity in its present form (hence the term *trans*humanist). While, of course, the incorporation of certain mechanical and computing components into ordinary humans, such as artificial hips or limbs or pacemakers can have a greatly beneficial effect on their recipients, the notion of taking things to the transhumanist extreme of "uploading" human "consciousness" to a machine strikes many (myself included) as, at the very least, farfetched in the extreme, and even downright creepy.

While advanced medical technologies certainly have their place, I, for one, want nothing to do with the transhumanist agenda. So—while not dismissing this field as pure whackery it is not one we will discuss further in this book. For an up-to-date (and humorous) discussion of this topic see Mark O'Connell's *To be a Machine* (2018). Ray Kurzweil's book *The Singularity is Near: When Humans Transcend Biology* (2005) also deals with this topic in a more general sense.

3.3   The Games People Play

> Denn, um es endlich auf einmal herauszusagen, der Mensch spielt nur, wo er in voller Bedeutung des Worts Mensch ist, und er ist nur da ganz Mensch, wo er spielt.
>
> For, to speak out once and for all, man only plays when in the full meaning of the word he is a man, and he is only completely a man when he plays.
> —Friedrich von Schiller (1795) Über die ästhetische Erziehung des Menschen in einer Reihe von Briefen.

Isaac Newton is said to have commented that many of his discoveries resulted from "random playing". The following excerpt from the introduction to Brian Burns' *Encyclopaedia of Games* (Burns 1998) illustrates eloquently the importance to humanity from the earliest times of games and gameplay.

> For thousands of years history has recorded that man has been a keen games player. The earliest recorded writings from ancient civilizations frequently refer to simple games such as Noughts and Crosses. As civilization has progressed so has the complexity of games. The earliest games played were probably those of the simple race variety, but many of these have

developed and matured through the centuries to become sophisticated modern games such as Chess and Shogi. References to games and game boards have frequently been found etched onto classical remains, such as Hadrian's Wall.

While there is no universally accepted definition of a game, a game is generally accepted to be a structured form of play. While games are generally enjoyed as a form of entertainment, there is also the genre of "serious" games—that is games devised for a particular practical function in order to enhance in some fashion the players mental and/or physical capabilities in some domain. Wittgenstein, in his *Philosophical Investigations* (2009), originally published posthumously in 1953 (Wittgenstein et al. 1953) wrote

> Consider...the activities that we call "games". I mean board-games, card-games, athletic games, and so on. What is common to them all?—Don't say: "they *must* have something in common, or they would not be called games...if you look at them you won't see something that is common to *all*, but similarities, affinities...

We may categorise games as purely mental requiring no physical activity (other than, for, example, the moving of a rook on a chess board, or mainly physical, such as rugby. We may also categorise games as games of skill, strategy, or luck; many games involve some combination of these three factors. We may then further subdivide games as board, card, dice, sports, video/computer, "serious", simulation type, etc.

While many games of skill and/or chance are now played as well as, or better than, the best human players (Go and Poker being the most recent to "fall"), it will still be many years before sports like soccer, or the ancient Irish game of hurling are played better by robot teams than by the best (or even mediocre) human players. But perhaps this is not so much the point. Games have, over the decades, been used as testing grounds and as benchmarks for AI research efforts. In many cases the real advances have not been so much in the ability of the AIE to beat the best human players, but that the associated research advances can then be applied to other domains outside of the game arena. For example, Watson, the IBM supercomputer that beat the best human competitors in the American quiz show *Jeopardy*, is now been successfully applied to medical diagnosis.

At a recent academic conference I had the opportunity to ask Minoru Asada, one of the founders of the RoboCup initiative in 1998 (Kitano and Asada 1998; Kitano et al. 1998), if he believed that the avowed aim of the initiative, that is that a team of humanoid soccer players would take on and beat the most recent winners of the FIFA world cup, would actually come about. After considering my question for a short while he answered (and I paraphrase) "I am not sure. But that is not really the main point". The main point being, of course, that the researches being made in this effort would hopefully contribute in a more general sense towards developments in advanced AI and robotics.

Charles Babbage, one of the earliest computer pioneers devised an Analytical Engine, essentially a mechanical general-purpose computer. While Babbage's main focus was undoubtedly on the application of his Analytical Engine to number crunching applications, he did envision its application to problems of a different nature, and decided after some reflection that game playing would be an appropriate domain.

> After much consideration I selected for my test the contrivance of a machine that should be able to play a game of purely intellectual skill successfully; such as tit-tat-to (sic), drafts, chess, &c.

Babbage in fact conceived of the construction of a 'tit-tat-to' playing machine consisting of two children playing the game in the company of a lamb and a cock, with the lab bleating in response to a loss, and the cock crowing in response to a win. Babbage considered the possibility that the construction of several such devices, exhibited in several places, might help fund the construction of his analytical engine. Upon further consideration, however, he concluded that

> to conduct the affair to a successful issue it would occupy so much of my own time to contrive and execute the machinery...that even if successful in point of pecuniary profit, it would be too late to avail myself of the money thus acquired to complete the Analytical Engine.

Tic-tac-toe, or noughts and crosses as it sometimes called, is often used to illustrate the application of artificial intelligence techniques. We will use it in Chap. 7 to illustrate both the minimax procedure for two-person adversarial game play and also to illustrate a learning technique called temporal difference reinforcement learning. A variant of the minimax procedure was at the heart of Deep Blue, the chess playing computer that defeated Garry Kasparov, the world chess champion in 1997. This we will discuss also. Other games/puzzles that we will use for illustrative purposes include the game of Nim, a river-crossing puzzle, and sliding-piece puzzles. In addition, we will look, in Chap. 8 at the application of evolutionary techniques to ball-kicking behaviour in a humanoid robot, an essential skill for robots competing in the RoboCup tournament. Finally, in Chap. 14 we will briefly discuss the AlphaGo Go playing program, and its successor AlphaZero—a learning system capable of playing chess, Go (Fig. 3.1), and shogi (Japanese chess) at superhuman levels.

Fig. 3.1 Two of the oldest board games in existence; Mancala (above) and Go (below). Avid Go enthusiasts may recognise the Go board configuration from the third game between Go world champion Lee Sedol and AlphaGo in which AlphaGo played the famous 37th move, described as "creative" and "unique" and "a move that no human would ever make". Photographs taken by the author

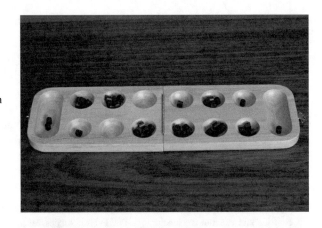

Chapter 4
Thought

Every act of thought may be looked at from two points of view. It may be regarded as a process in time, that is, as a mere psychological event, or as a meaning. As a process in time, it is a state of consciousness among other such states to which it is related and by reference to which it may be explained. As a meaning, it is the expression of the relation of subject to object, the expression of which relation gives it its significance as an act of knowledge. Neither of these aspects of thought can, of course, be neglected; a timeless act of thought is as much a nonentity as a meaningless act of thought. But, on the other hand, the two aspects must not be confused; thought as a process in time is something quite different from thought as a meaning.

—*Gustavus Watts Cunningham (1910).* Thought and reality in Hegel's system

We think in eternity, but we move slowly through time.
—*Oscar Wilde, De Profundus*

4.1 "Pure" Thought

In general, we find it hard to conceive of the notion of thought without reference to human thinking and human thought processes. Is it really possible to conceive of thought, and of a train of thinking, *outside* of the human experience? In a way, thought in an abstract sense— "disembodied thought"? What, indeed, are the core essential features of thought and the thinking process as conceived outside of our physical selves?

In fact, many scholars and philosophers would question the notion (or existence) of "pure" thought, that is a thinking process divorced from human, animal, alien, or machine. In this context his chapter on "pure" thought will remain short (on fact the shortest in this book) not least because of the inherent difficulty of the subject matter.

In his controversial and groundbreaking book, *The origin of consciousness in the breakdown of the bicameral mind* (1976) the psychologist Julian Jaynes argued that consciousness is not necessary for thinking. Using an example of a person picking

© Springer Nature Switzerland AG 2020
M. Eaton, *Computers, People, and Thought*,
https://doi.org/10.1007/978-3-030-55300-5_4

up, with eyes partially closed, two objects of unequal weight, and given the task of judging which is the heavier, while one is conscious of the feel and shape of each of the two objects the act of judging which is the heavier (judgment thinking) he would claim is not a conscious act—"it is just somehow given to you by your nervous system". Jaynes further argues (Jaynes 1976)

> So we arrive at the position that the actual process of thinking, so usually thought to be the very life of consciousness, is not conscious at all and that only its preparation, its materials, and its end result are consciously perceived.

In fact the central thesis of *The origin of consciousness in the breakdown of the bicameral mind* is that before around 1000 BC humankind was not in fact conscious in the way we understand it today but instead acted on "voices" in their head originating in the right hemispheres of their brains.

4.2   Thought as the Shaper of Reality Itself

The prominent Prussian philosopher and scientist Immanuel Kant (1724–1804) held the view that thought, rather than being a reflection of reality was in fact the shaper of reality itself. This was one of the core tenets of his "Copernican Revolution". As Schwyzer (1973) puts it

> The nature of things we think about, things conceptualizable, is determined by the nature of thinking about things, of conceptualizing things. That is the Copernican Revolution.

and further

> Kant and Wittgenstein are together in this: the necessities in our thought about things, necessities which constitute the nature of the things we think about, have their source not in the things themselves, but in the nature of our thinking.

Ludwig Wittgenstein (1889–1951), referred to in the above quote was an Austrian philosopher, regarded by many scholars as the most important philosopher of the twentieth century. His posthumously published book *Philosophical Investigations* (1953) is widely regarded as masterpiece in philosophical writing. Early in the *Investigations* he writes

> Thinking is surrounded by a nimbus [halo]. —Its essence, logic, presents an order: namely the a priori order of the world: that is, the order of *possibilities*, which the world and thinking have in common.

and further on in the *Investigations*

> Well, what does one call "thinking"? What has one learnt to use this word for? — If I say I've thought—need I always be right? —What *kind* of mistake is there room for here? Are there circumstances in which one would ask, "Was what I was doing then really thinking; aren't I making a mistake?" Suppose someone takes a measurement in the middle of a train of thought: has he interrupted the thinking if he doesn't say anything to himself while measuring?

I think it should be clear from the above brief discussion that the subject of "pure" thought, while fascinating is a difficult and complex one from a conceptual perspective. The interested reader (perhaps of a philosophical bent) is encouraged to follow further the references given, and to form their own perspective on this both intriguing and challenging topic.

Part II
The Synergies

Chapter 5
Computers and People

5.1 Some Important Historical Figures

As will become abundantly clear after a short examination of the contents of this book, the focus of this book lies squarely on principles rather than personalities. This text is not about historical figures, or the general history of computing or of "AI", whatever that might mean. However there are certain figures in the history of this field—including Charles Babbage, Claude Shannon, Alan Turing, John von Neumann, and others, whose contributions have been so important and significant that to omit an appropriate brief acknowledgement, including a short synopsis of their lives, and of their major scientific contributions, would not seem entirely correct. We might yet have arrived at the same result in terms of the development of Artificially Intelligent Entities (AIEs), but the path taken could well have been significantly different, and, perhaps, much longer.

However, be aware that this is a personal list, selected by the author and that many, many, more people made major contributions to the field; those listed happen to have particular relevance to the focus of this text. They are listed here in chronological order of their birth.

5.1.1 Charles Babbage

Charles Babbage (1791–1871) may rightly be considered the world's first major computer pioneer and a true intellectual. A Cambridge professor, Babbage had wide ranging interests, from astronomy to statistics, and philosophy. However, it is probably from his work as a computer pioneer that he is best known today.

Following initial work on a device he called the *Difference Engine*, designed to tabulate mathematical functions, he turned his attention to a far more ambitious project—the *Analytical Engine*, essentially a general purpose mechanical digital

© Springer Nature Switzerland AG 2020
M. Eaton, *Computers, People, and Thought*,
https://doi.org/10.1007/978-3-030-55300-5_5

computer. Much of Babbage's work is described in a series of notebooks, or sketchbooks together with many detailed drawings of the Analytical Engine. Interestingly, while using the decimal system for storage and calculation it is clear that Babbage also considered the relative merits of other bases (Wilkes 1977)

> Early in the original enquiry I had examined the relative value of various bases of notation in arithmetic, 10, 12, 16, 20, etc., and had contrived a carriage for an engine with base 100 so that each figure wheel would have contained two places of figures. These were given up for reasons then stated in other papers. I had also tried bases less than 10 as 5, 4, 3, 2, but these were rejected on account of the great multitudes of wheels required.

Unfortunately, although a true computer pioneer, Babbage never put his inventions and discoveries into a readily digestible form and, as Wilkes put it "everything that he discovered had to be re-discovered later" (Wilkes 1977).

Babbage also had a very human side, as is made clear in his book *Passages from the life of a Philosopher* (1864). In particular he was clearly not an advocate of street music, and its associated "instruments of torture", which, he claimed

> robs the industrious man of his time; it annoys the musical man by its intolerable badness; it irritates the invalid...and it destroys the time and the energies of all the intellectual classes of society by its continual interruptions of their pursuits.

Babbage further divided musical performers into the following broad categories (Babbage 1864)

Musicians	*Instruments*
Italians	Organs
Germans	Brass bands
Natives of India	Tom-toms
English	Brass bands, fiddles, &c.
The lowest class of clubs	Bands with double drum.

5.1.2 George Boole

George Boole (1815–1864) laid the foundations for Boolean algebra, which forms the basis of operation of almost all modern digital computers. His most famous work *An Investigation of the Laws of Thought on Which are Founded the Mathematical Theories of Logic and Probabilities* (Boole 1854) followed on from an earlier work, *The Mathematical Analysis of Logic* (Boole 1847) "but its methods are more general, and its range of applications far wider" (Boole 1854). Referencing Aristotle and many other earlier writers Boole sought to investigate and apply mathematical rigour to the process of reasoning and other areas "concerning the nature and constitution of the human mind".

Boole's work was purely theoretical—unlike Babbage he did not seek to give physical realisation to the concepts he explored so thoroughly. Born in England, Boole was appointed the first professor of mathematics in the then recently established Queen's College, Cork (now called University College Cork, Ireland).

5.1.3 Norbert Wiener

The new industrial revolution is a two-edged sword. It may be used for the benefit of humanity, but only if humanity survives long enough to enter a period in which such a benefit is possible. It may also be used to destroy humanity, and if it is not used intelligently it can go very far in that direction.

—Norbert Wiener (1954). *The Human Use of Human Beings*. Houghton Mifflin Company, New York.

Norbert Wiener (1894–1964) was the founding father of the Cybernetics field, closely allied to the Artificial Intelligence movement, and a renowned mathematician. In common with Edmund Berkeley, Wiener had profound reservations about the potential applications of advanced technology. After the Second World War Wiener refused to work for the military, or to accept government funding.

5.1.4 John von Neumann

Technology—like science—is neutral all through, providing only means of control applicable to any purpose, indifferent to all.... For progress there is no cure. Any attempt to find automatically safe channels for the present explosive variety of progress must lead to frustration. The only safety possible is relative, and it lies in an intelligent exercise of day-to-day judgment.

—John von Neumann "Can We Survive Technology?", Fortune, June 1955 issue.

John (Johnny) von Neumann (1903–1957) was a man of astonishing intellectual ability by any measure. His contributions to human knowledge ranged from the fields of pure mathematics including quantum theory and mechanics, ergodic theory and operator theory, to issues of theoretical hydrodynamics, meteorology, and various aspects of automata theory, founding the field of cellular automata. He also made highly significant contributions to game theory and to mathematical economics.

His work on theoretical hydrodynamics led him into being involved in the Manhattan project and the creation of the atomic bomb, also making significant technical contributions to the development of the hydrogen bomb, and was in favour of the U.S. adopting a strong military position. He was also a member of the U.S. Atomic energy commission. von Neumann made many significant contributions to the development of modern computing, and to the relationship between brains and computing. This was the subject of his book *The Computer and the Brain* (1957), one of his last published works.

5.1.5   Edmund Berkeley

> It would be reasonable for every person, computer scientist or not, who sees these dangers clearly to devote a substantial portion of his time, energy, and resources to helping to prevent the logical lethal consequences of computers plus nuclear weapons plus rocket power.
> —Edmund Berkeley (1961). *Giant Brains,* 2nd Edition Wiley & Sons.

Edmund Berkeley (1909–1988) was one of the driving forces in the early history of computation, and was one of the co-founders of the *Association for Computing Machinery (ACM)* in 1947, one of the foremost worldwide computing organisations, and which presents the annual Turing award, regarded as the Nobel Prize for computer science.

A committed educationalist, Berkeley wrote the first major text popularising computation for the masses, entitled *Giant Brains, or Machines that Think*, originally published in 1949. Berkeley was also acutely aware of the great potential dangers associated with the unbridled application of technological advances by those without a social conscience.

5.1.6   Alan Turing

> As soon as one can see the cause and effect working themselves out in the brain, one regards it as not being thinking, but as unimaginative donkey-work. From this point of view one might be tempted to define thinking as consisting of "those mental processes we don't understand". If this is right then to make a thinking machine is to make one which does interesting things, without our really understanding how it is done.
> —Alan Turing (1952) Transcript from radio broadcast, from Alan Turing Archives

Alan Turing (1912–1954) is regarded by many as the father of Artificial Intelligence, although the term was originally coined by John McCarthy (and others) in 1955 in the proposal for the 1956 Dartmouth conference on this topic (McCarthy et al. 1955). Like von Neumann, he was active in many areas of mathematics and computing. He also worked in cryptography and in breaking the Enigma codes used in the Second World War by the Germans. He is perhaps best known in popular literature for the so-called *Turing Test* designed as a test of machine intelligence (see the next section). Unlike von Neumann, Turing was convinced that A^2IEs equalling and even surpassing human intelligence could be constructed.

Turing was an excellent long-distance runner and could have been a marathon contender for the Olympic games were it not for injury. Alan Turing died in 1954 at the age of 41, some say by suicide, but many claim as a result of a tragic accident. Despite his premature death his legacy lives on to the present day (see Fig. 5.1).

Fig. 5.1 Photographs taken by the author on a wet October afternoon at the Alan Turing memorial statue outside Manchester University during the 2013 IEEE Systems, Man, and Cybernetics (SMC) Conference in Manchester, England

5.1.7 Claude Shannon

Claude Shannon (1916–2001) is perhaps best known for his 1948 article "A Mathematical Theory of Communication" published in the *Bell System Technical*

Journal, in which he laid out the basic principles for the efficient transmission of digital information (Shannon 1948). However a decade earlier, in 1938, he published a paper entitled "A symbolic analysis of relay and switching circuits" (which was his Master's thesis from MIT a year earlier), in which he demonstrated how logic circuits based on Boole's abstract analyses from the previous century could be implemented using switching circuits based on electromechanical relays (Shannon 1938). In doing so he paved the way for the engineering implementation of the modern digital computer.

5.1.8 Marvin Minsky

Once the computers got control we might never get it back. We would survive at their sufferance. If we're lucky, they might decide to keep us as pets.
—Marvin Minsky (1970) Life Magazine, Nov 20th, 1970, p. 68

Marvin Minsky (1927–2016) was a highly influential figure in the field of AI from its "inception" in 1956 (he was a participant in the Dartmouth conference) to his death in 2016. He was the author of several influential textbooks in the field and was one of the early recipients of the Turing Award in 1969. He was the fourth winner of this annual award (it started in 1966) and the first researcher working specifically in the AI field to win the award. John McCarthy (one of the originators of the term "Artificial Intelligence") won the award 2 years later in 1971.

5.1.9 Rodney (Rod) Brooks

Technology has unintended consequences. Sometimes they are large and tumultuous. It is often well worth the trouble of trying to figure them out ahead of time.

...and flying cars—forget about 'em.
—Rodney Brooks (2017). Robotic cars won't understand us, and we won't cut them much slack. IEEE Spectrum, 54(8), 34–51.

Rodney (Rod) Brooks (1954–) is a roboticist, entrepreneur, and former Professor and director of the MIT Computer Science and Artificial Intelligence Laboratory (CSAIL). He is probably best known for his early work on the *subsumption architecture* consisting of networks of augmented finite state machines passing messages between each other in a layered fashion. In this system, unlike in traditional robot control architectures there is no central control locus. Rod is also known for his work on the COG upper-body humanoid robot developed in MIT in the 1990s.

5.2   Asimov's Laws and Turing's Test

The Three Laws of Robotics

1—A robot may not injure a human being or, through inaction, allow a human being to come to harm.

2—A robot must obey orders given it by human beings except where such orders would conflict with the First Law.

3—A robot must protect its own existence as long as such protection does not conflict with the First or Second Law.

HANDBOOK OF ROBOTICS
   56th EDITION 2058 A.D.
   —Isaac Asimov (1950), *I, Robot*. Gnome Press.

I believe that in about fifty years' time it will be possible to programme computers, with a storage capacity of about 10^9, to make them play the imitation game so well that an average interrogator will not have more than 70 per cent, chance of making the right identification after five minutes of questioning. The original question, "Can machines think !" I believe to be too meaningless to deserve discussion. Nevertheless I believe that at the end of the century the use of words and general educated opinion will have altered so much that one will be able to speak of machines thinking without expecting to be contradicted. I believe further that no useful purpose is served by concealing these beliefs.
   —Alan Turing (1950) "Computing machinery and intelligence". *Mind*, 59, 433–460.

By the judicious combination of Asimov's "three laws" with the "Turing test" we have a way of not only testing whether we have created human-level intelligence, but also a method of controlling it. So goes the theory.

However right from the beginning Asimov acknowledged the numerous problematic issues associated with his three laws—and took advantage of these inconsistencies to construct many story lines.

For example, in *Runaround*, the short story where the laws were first introduced to the world in 1942 (this story was later reproduced in the anthology *I Robot* (1950)), the robot in question, Speedy, has been sent to retrieve life-saving selenium from the surface of a hostile planet. However, because of the advanced and costly nature of the robot, the strength attached to the third law has been increased to reduce the likelihood of the robot coming to harm. So, when it goes to retrieve the selenium, and it detects a dangerous gas in its vicinity, because the order to retrieve the selenium was not issued in a sufficiently authoritative fashion, at a certain point distant from the selenium laws 2 and 3 perfectly balance, and instead of retrieving the life-saving substance it instead circles it, in equilibrium between the potential of the two laws—the *runaround* of the title. The situation is only retrieved when one of the human protagonists puts himself in imminent danger within Speedy's reach, when law 1 (not allowing a human to come to harm) comes into play, overriding the potential of the other two laws; the endangered scientist is transferred safely back to base, where Speedy is ordered to retrieve the selenium at any cost, and a happy ending ensues with the robot apologising profusely "for the runaround he gave us".

In a sense, Asimov's "three laws" can be consigned to the same piece of history as the Turing test: famous concepts, highly influential, but ultimately outdated for application to present day technologies in their original formulations. So, while undoubtedly of great historical importance, Asimov's laws and Turing's test are ultimately not of great use in the present day in their original form. This is not to underestimate their overall importance, however, as they still, today, serve to fuel debates over the benchmarking/evaluation (Turing) and the safety (Asimov) of future advanced artificial intelligent entities (A^2IEs).

5.3 Bio-inspired Computation and Beyond

In this section we will address the general areas of bio-inspired, nature-inspired, and metaphor-inspired systems. People (humankind) are generally accepted to have developed, at least in part, by a natural evolutionary process, arising from the primordial slime. Evolutionary algorithms are an attempt to recreate, using computers, aspects of this natural evolutionary process in an artificial context. Evolutionary algorithms and evolutionary computation fall into the category of what are generally termed *bio-inspired* systems, which include other paradigms inspired by biological systems, such as artificial neural networks.

Bio-inspired computation in turn falls under the more general (and more recent) category of nature-inspired systems; that is, any computational process inspired by any aspect of the natural world—biological or otherwise. Finally, nature-inspired systems, in turn, fall under the very general category of metaphor-based, or metaphor-inspired systems. Metaphor-inspired systems have come in for some bad press in recent times—not always, it may be argued, completely unjustified. This is an issue we will return to briefly again at the end of this section.

5.3.1 Evolutionary Algorithms

If nature-inspired algorithms/metaheuristics are algorithms inspired by natural processes and forces, there is not a much more powerful force in nature, certainly from a biological perspective, than natural evolution. Evolutionary algorithms can trace some of their early inspiration to Alan Turing who speculated in his highly influential 1950 paper "Computing Machinery and Intelligence" on the development of machine intelligence by starting with a mechanical "child brain" that could be "easily programmed" (Turing 1950). Turing had an interesting variant on natural selection, which he considered too slow; this consisted in the replacement of natural selection by the "judgement of the experimenter".

> We cannot expect to find a good child machine at the first attempt. One must experiment with teaching one such machine and see how well it learns. One can then try another and see

if it is better or worse. There is an obvious connection between this process and evolution, by the identifications

Structure of the child machine = Hereditary material

Changes   „   „   = Mutations

Natural selection   = Judgment of the experimenter

One may hope, however, that this process will be more expeditious than evolution. The survival of the fittest is a slow method for measuring advantages. The experimenter, by the exercise of intelligence, should be able to speed it up.

This idea can be considered analogous to the modern paradigm of *interactive evolutionary computation*, as proposed originally by Dawkins (1986) in his influential book and expounded further by Takagi (2001). Friedberg and his colleagues came up with some of the earliest practical attempts to apply evolutionary techniques in the evolution of computer programs (Friedberg 1958; Friedberg et al. 1959), and as such may be seen as an early forerunner of Holland's influential work on *genetic algorithms* and *classifier systems* (Holland 1975), which is generally regarded as the first major work heralding evolutionary algorithms as a separate important research field (although the term *evolutionary algorithm* was not in general use at this time). We will discuss the operation of a basic genetic algorithm in Chap. 7.

However it should be noted that Friedberg et al. did not uses the term "evolution" in their description of their work, and their work did come in for some criticism, most notably from Marvin Minsky in his influential 1971 article "Steps towards Artificial Intelligence" (also reproduced in the anthology *Computers and Thought*) where their work was described as a "comparative failure" (Feigenbaum and Feldman 1963).

Many other lines of research followed on, based broadly on the principles of the evolutionary process in nature, including *evolutionary programming* (EP, Fogel et al. 1966), *evolutionary strategies* (ES, Rechenberg 1973), *genetic algorithms* (GA, Holland 1975), *genetic programming* (GP, Koza 1992), *grammatical evolution* (GE, O'Neill 2001), *covariance matrix adaptation* (CMA, Hansen and Ostermeier 2001), *neuroevolution of augmenting technologies* (NEAT, Stanley and Miikkulainen 2002), and the *nondominated sorting genetic algorithm II* (NSGA-II, Deb et al. 2002).

Clearly, therefore, the field of evolutionary algorithms has a long and impressive research pedigree with a proven record of research results. Other well-established nature-inspired algorithms (sometimes referred to as *natural computing algorithms*), which we will briefly look at now, include *simulated annealing* (Kirkpatrick et al. 1983), *ant colony optimisation* (Colomi et al. 1991), and *particle swarm optimisation* (Kennedy and Eberhart 1995).

5.3.2   Simulated Annealing

While genetic and other evolutionary algorithms draw inspiration from evolutionary biology, simulated annealing has its roots in statistical physics. *Annealing* is the term given to the procedure used to harden glass and metals by subjecting them initially to a very high heat, and then gradually allowing them to cool, thus in the process reaching a stable crystalline state.

Simulated annealing uses a procedure analogous to this in order to escape from local minima (the main drawback of *hill-climbing* search, which, as its name implies, simply involves moving our search in small increments in the direction that brings us closest to the solution at each iteration of the algorithm) by also allowing the search algorithm to choose a random (rather than the assumed best) move at each iteration. If this move results in an improvement over the previous state then it is accepted. If, on the other hand, the move results in a worse state the move may still be accepted with probability p. This probability is related to how 'bad' the move is perceived to be—i.e. a very bad move would be taken very rarely, while a move that only slightly disimproves the current situation would be taken with a relatively high probability.

As in the physical annealing process, the overall probability of taking a 'bad' move is relatively high initially (associated with high 'temperature' T) and this probability reduces gradually as the temperature is decreased. If T is reduced gradually enough, simulated annealing can be demonstrated to find a global (rather than a local) optimum with high probability.

5.3.3   Particle Swarm Optimisation

Particle swarm optimisation (PSO) is based on the idea of simulating in graphical form the behaviour of flocking birds. It was originally designed to emulate the behaviour of this elegant and unpredictable movement of birds. It then developed into an optimisation algorithm based on particles being "flown" through hyper-dimensional space, where individual particles are influenced by the behaviour of other, successful, particles in their neighbourhood. Thus, the driving force behind PSO is social interaction, with particles striving to be more similar to their "superior" neighbours.

Like evolutionary algorithms, particle swarm optimisation does a parallel search through the search space using a range of potential solutions (*chromosomes* in the EA case, *particles* in PSO). However, unlike EAs the final solution is arrived at by *cooperation* between these candidate solutions rather than *competition*. PSO has been successfully applied in a number of domains, most notably in the optimisation of non-linear functions (Engelbrecht 2002).

5.3.4 Ant Colony Optimisation

Ant colony optimisation is a search technique based on the behaviour of ants foraging for food using pheromones to guide their search. These pheromones are dropped by ants on their food-foraging expeditions, and in selecting paths to follow ants follow paths with the relatively greater pheromone concentration. The strength of pheromones deposited reduces over time. A variety of different ACO algorithms have been developed to address different problem domains. In general, the ACO approach is suited to finding paths through complex landscapes that are dynamically changing. This is because the ACO will generally maintain alternative paths through the graph in addition to the shortest one and is able to quickly change to one of these alternative paths if the best path is suddenly made unavailable.

5.3.5 A Wild Frontier

Having worked for a significant portion of my research career in the areas of evolutionary algorithms and neural networks, both of which paradigms derive direct inspiration from natural processes. I was somewhat intrigued in the course of research for this book to learn of the number of algorithms today deriving inspiration from biological systems, from natural processes in general, or even such an improbable area as mimicking the improvisation of human musicians. Most of these new algorithms (sometimes referred to as *metaheuristics*) were invented in the last decade or so, with some exceptions. So, let us now take a brief foray into the wild (and sometimes wacky) world of metaphor-inspired algorithms.

5.3.5.1 The Metaheuristic Menagerie

This is where things start to get more interesting (and even, perhaps some might say, a little strange). While all of the algorithms outlined in the previous sections (EAs, SA, PSO, and ACO) together with their different variants have a proven track record, over the last 20 years or so a plethora of different nature-inspired or metaphor-inspired algorithms have come to the fore, a number of which have attracted critical attention. This is generally either because of the dubious nature of the research results presented, or because on closer inspection these algorithms turn out to be just variants on the various established natural computing algorithms we have just discussed. This is not, of course, to say that any novel metaphor-inspired algorithm should be dismissed out of hand, just that a little caution is advised.

Harmony search (Geem et al. 2001) is in a sense the granddaddy of these metaheuristics, widely referenced and used—the original article is cited over 3500 times on Google Scholar at the time of writing. The principle of operation of harmony search is evocatively and concisely described by Yang (2009):

Harmony search is a music-based metaheuristic optimization algorithm. It was inspired by the observation that the aim of music is to search for a perfect state of harmony. The effort to find the harmony in music is analogous to find the optimality in an optimization process. In other words, a jazz musician's improvisation process can be compared to the search process in optimization. On one hand, the perfectly pleasing harmony is determined by the audio aesthetic standard. A musician always intends to produce a piece of music with perfect harmony. On the other hand, an optimal solution to an optimization problem should be the best solution available to the problem under the given objectives and limited by constraints. Both processes intend to produce the best or optimum.

Other metaphor-based methodologies include *bee colony optimization* (BCO, Teodorović et al. 2006), the *big bang–big crunch algorithm* (BB-BC Erol and Eksin 2006), the *blind naked mole-rats algorithm* (BNMR, Taherdangkoo et al. 2013), the *black hole algorithm* (Hatamlou 2013), the *interior search algorithm* (ISA, Gandomi 2014), the *chicken swarm algorithm* (CSO, Meng et al. 2014), the *elephant search algorithm* (ESO, Deb et al. 2015), the *whale optimization algorithm* (WOA, Mirjalili and Lewis 2016), and the *African buffalo optimization algorithm* (ABO, Odili et al. 2015).

While it is not proposed to go into any major detail of these individual algorithms—the reader is encouraged to refer to the original articles if interested—perhaps a couple of short quotes will serve to give a flavour of the ideas presented.

Regarding the interior search algorithm (ISA) we have (Gandomi 2014):

The proposed ISA is inspired by interior design and decoration....The ISA takes into account the aesthetic techniques commonly used for interior design and decoration to investigate global optimization problems, therefore, it can also be called aesthetic search algorithm.

And for the chicken swarm optimization (CSO, Meng et al. 2014)

A new bio-inspired algorithm, Chicken Swarm Optimization (CSO), is proposed for optimization applications. Mimicking the hierarchal order in the chicken swarm and the behaviors of the chicken swarm, including roosters, hens and chicks, CSO can efficiently extract the chickens' swarm intelligence to optimize problems.

A recent proposal (Odili and Kahar 2016) is for African buffalo optimisation (ABO)

This paper proposes the African Buffalo Optimization (ABO) which is a new metaheuristic algorithm that is derived from careful observation of the African buffalos, a species of wild cows, in the African forests and savannahs...In ABO, our interest is in how the buffalos are able to organize themselves in searching the solution space with two basic modes of communications, that is, the alarm "waaa" sounds to indicate the presence of dangers or lack of good grazing fields and, as such, asking the animals to explore other locations that may hold greater promise. On the other hand, the alert "maaa" sound is used to indicate favourable grazing area and is an indication to the animals to stay on to exploit the available resources.

Finally, while a wide variety of nature inspired algorithms exist, one that particularly caught my attention (partly because of my fondness for these feline creatures), was the cat swarm optimisation (CSO) algorithm (Chu et al. 2006) and its close cousin the parallel cat swarm optimisation algorithm (PCSO) (Tsai et al. 2008). We can get a flavour of the *modus operandi* of the PCSO with a short quote from the latter paper.

The process of CSO is described as follows:

1. Create N cats in the process.

2. Randomly sprinkle the cats into the M-dimensional solution space and randomly assign values, which are in-range of the maximum velocity, to the velocities of every cat. Then haphazardly pick number of cats and set them into tracing mode according to MR, and the others set into seeking mode.

3. Evaluate the fitness value of each cat by applying the positions of cats into the fitness function, which represents the criteria of our goal, and keep the best cat into the memory.

In fact, the whole field of cat swarm optimisation has matured to the extent that there is a review paper dedicated to this topic alone (Tsai and Istanda 2013).

5.3.5.2 Combining Algorithms

It is quite common to combine different nature inspired algorithms in order to obtain the "best of both worlds". For example, this is an abstract from a recent article combining features from the "bat algorithm" and the "artificial bee colony algorithm" (Pan et al. 2014):

In this paper, a hybrid between Bat algorithm (BA) and Artificial Bee Colony (ABC) with a communication strategy is proposed for solving numerical optimization problems. The several worst individual of Bats in BA will be replaced with the better artificial agents in ABC algorithm after running every *Ri* iterations, and on the contrary, the poorer agents of ABC will be replacing with the better individual of BA. The proposed communication strategy provides the information flow for the bats to communicate in Bat algorithm with the agents in ABC algorithm.

5.3.5.3 A Gentle Critique

One issue that arises, and which makes many of these articles difficult to comprehend at first reading, can be the use of very metaphor-specific language where concepts already familiar to the search and optimisation community are renamed to conform with the particular metaphor under discussion. Also, while there are undoubtedly new ideas and exciting new concepts contained within the large body of metaphor-inspired algorithms, in some (many?) cases there is an element of "rediscovering" existing concepts under the cloak of complex metaphor-based language.

One methodology which has come under particular scrutiny (perhaps unfairly) is the harmony search algorithm, discussed earlier (Weyland 2010, 2015; Geem 2010). While we do not intend to engage in direct criticism of any of the heuristic methods mentioned in this section, certain questions do undoubtedly need to be answered.

Kenneth Sörensen (2015) succinctly puts this case in his recent widely quoted article "Metaheuristics—the metaphor exposed":

> ..."novel" metaheuristics based on new metaphors should be avoided if they cannot demonstrate a contribution to the field. To stress the point: renaming existing concepts does not count as a contribution. Even though methods may be called "novel" by their originator, many present no new ideas, except for the occasional marginal variant of an already existing method. Moreover, these methods take up the space of truly innovative ideas and research, for example in the analysis of existing heuristics. Because these methods invariably change the vocabulary, they are difficult to understand. Combined with the fact that the authors of these methods usually neglect to properly position "their" method in the metaheuristics literature, such methods present a loss of time and a step backward rather than forward.

5.4 Embodied Intelligence

We can view embodied intelligence as intelligent agents with the ability to materially affect their external environment. Humans (people) can be viewed as a natural manifestation of embodied intelligence, whereas certain advanced humanoid robots might be seen as an artificial manifestation of such intelligence. Of course, all intelligence is embodied to a certain degree—unless we are to talk about potential spiritual entities, and that is well outside the scope of this book. However, it should be noted that researchers as eminent as Alan Turing did not discount the possibility of the interference of elements outside of the realm of conventional science—in particular he considered the possibility that telepathic abilities might give humans an advantage in a version of the imitation game to the extent that he suggested the possibility of the construction of a "telepathy-proof" room to avoid the possibility of any such occurrence (Turing 1950). As Turing remarked "With E.S.P. (Extra Sensory Perception) anything may happen."

On a more practical level, your desktop machine probably has a hard disk drive with magnetic platters spinning at high speed creating heat and altering the characteristics of its immediate environment. Also, your computer screen generates light patterns which your eyes interpret, etc. But it cannot be said that your desktop machine has a major material effect on its external environment. For example it can't hand you a cup of coffee, or rearrange your desk into a more aesthetically pleasing configuration, or open the window in your office to let in fresh air.

Add robotic manipulators and/or some means of locomotion however and the picture changes. We now have an embodied intelligent agent—assuming, of course, that it can now perform some tasks that we would deem "intelligent" using these effectors, such as playing a game of chess by moving pieces on a physical board, finding its way through a physical maze, etc.

5.4.1 Humanoid Robots

A broad definition of a humanoid robot is a robot that is humanlike to a greater or lesser extent in either its behaviour, its morphology (shape), or both (Eaton 2015). However, in the general public perception of humanoid robots (which, it has to be said has been formed in no small part by films such as The Terminator, RoboCop and Ex Machina) a humanoid robot is a roughly adult-sized robot with two arms, two legs, etc. and which is capable of carrying out reasonable conversations and in acting in most ways broadly similar to (or in many cases more effectively than) a human.

No such humanoid robots exist in the real world today. The vast majority of robots that we might characterise as humanoid are a lot smaller than an adult human and are mainly used for research or entertainment purposes. Two such robots are shown in Fig. 5.2 below: the Darwin-Mini and Bioloid (Premium) humanoids, both from the Korean company Robotis. Also pictured in the following two figures are two mid-sized robots, the Nao and the iCub humanoids, both used extensively in robotic research programs (Fig. 5.3), and two adult-sized robots, the REEM-C and TEO humanoids (Fig. 5.4).

Probably the most advanced humanoid in production today that is available for purchase is the Honda ASIMO (Advanced Step in Innovative Mobility), which has been in development since its introduction in 2000, following Honda's long

Fig. 5.2 Two mini-sized robots, both from the Korean company Robotis. On the left is the Bioloid humanoid (which comes in kit form) standing 40 cm high, and on the right the Darwin-Mini which is 27 cm tall. Each robot retails for less than around 1000 €

Fig. 5.3 Two mid-sized (toddler-child size) robots; on the left is the Aldebaran Robotics Nao humanoid which is the robot used in the RoboCup Standard Platform League (SPL) robot soccer tournament, and which is 27 cm tall. On the right is the iCub open source humanoid which is 100 cm high. Prices for these platforms vary, around 10,000 € for the Nao, and a hefty 250,000 € for the iCub humanoid

Fig. 5.4 Examples of two full-sized (adult-size) humanoids, both from Spain. On the left is the 165 cm tall REEM-C humanoid from PAL robotics, based in Barcelona; on the right is the TEO humanoid which stands approx. 150 cm, and is under development by the Carlos III University of Madrid. All photographs taken by the author

involvement in humanoid robotics development going back to the 1980s (Hirose and Ogawa 2007). ASIMO can be purchased for a price in the region of a mere 2,000,000 € ($2,500,000). Standing 130 cm tall, ASIMO was judged to be the ideal height to act as a mobility assistant, allowing it to interact with everyday devices, serving at table, operating door handles, and turning switches on and off.

Inspired by ASIMO, Honda have also recently unveiled a new experimental legged robot designed for inspection and disaster response in plants, including nuclear plants such as the Fukushima Daiichi nuclear power plant, damaged by the 2011 Great East Japan earthquake (東日本大震災,). This humanoid, titled E2-DR (experimental robot type 2 for disaster response) contains a number of innovative features including a novel cooling system, and the ability to manoeuvre on stairs and different ladder types and to operate in narrow free spaces (Yoshiike et al. 2017).

Of course, a major motivation for research in the area of humanoid robotics is that by developing a robot in broadly the same form factor as a human, this robot should, theoretically, be able to operate in any environment that a human operates, thereby replacing or assisting humans in any such environment. This is of particular interest in three types of labour environments, those that are *dirty*, *dangerous*, or *dull* (the three Ds). An example of a dirty environment might be in the area of household rubbish removal or manipulation; work in a radioactive environment such as in the aftermath of the Great East Japan Earthquake is clearly dangerous for humans; and tedious factory line assembly line operation is an example of dull work. Of course, in the future, with the rapid advances we are seeing in AIE technology, together with impressive advances in the robotics field, in the long term it may be that virtually all jobs currently performed by humans may be performed equally well or better by humanoid robots. This is a topic we will return to in more detail in Part III.

Chapter 6
People and Thought

> *. . .although we have focused on the future of intelligence in this book, the future of consciousness is even more important, since that's what enables meaning. Philosophers like to [contrast]* sapience *(the ability to think intelligently) with* sentience *(the ability to subjectively experience qualia). We humans have built our identity on being Homo sapiens, the smartest entities around. As we prepare to be humbled by ever smarter machines, I suggest that we rebrand ourselves as* Homo sentiens!
> —*Max Tegmark* Life 3.0

6.1   Human Consciousness

The concepts of "thought" and of "consciousness" both belong to the same ephemeral realm, which is perhaps not so helpful when talking about the construction of real physical artefacts. However, if we are to look at thought as *the conscious manifestation of one's ideas and/or intentions*, then from this perspective it is clear that consciousness is a prerequisite for thought.

In a certain sense it might even be claimed that many aspects of reality cannot be said to exist without some conscious entity perceiving them. So—is it possible that the entire universe could not have come into existence originally because of the lack of conscious entities (certainly at the earlier stages)—unless we presuppose the prior existence of a conscious entity (to "will" it into existence)?

Consciousness is indeed a fragile thing. In everyday life, through chores and routines consciousness appears to flow and vanish as if unreal. Then, at other times of depression and/or melancholia the conscious knowledge of one's existence seems to bear down like a heavy weight. It is a peculiar characteristic of the human condition that the very acknowledgement and appreciation of one's conscious self can be unhelpful in daily life. Is consciousness then simply a mechanism for habit formation? I think not. Let us explore further.

The solipsist perspective is that only you yourself (me) can be truly looked at as a conscious being. Indeed, the words written here and your perception thereof of my conscious aspect are in one sense imaginary. However, even accepting the solipsist

© Springer Nature Switzerland AG 2020
M. Eaton, *Computers, People, and Thought*,
https://doi.org/10.1007/978-3-030-55300-5_6

viewpoint (which you may view as a little extreme) still does not prevent one from being affected by and having thought trains set in motion as a result of reading this text.

Relaxing our solipsist perspective for the time being, we survey our fellow human beings. We will accept then that they are also conscious entities (or at least our close acquaintances are). What about the cat next door? She seems to have many of the elements of consciousness; the ability to feel pain, apparent happiness, etc.—perhaps she also has limited consciousness? But let us pause for a second here. A lot of the problem seems to stem from the vagueness of our definition of consciousness; indeed, no concrete definition has, so far, been offered. Is consciousness the ability to feel emotions such as anger and happiness? What, indeed, do we mean by "feel"? This is an interesting question, and, of course, emotions are not to be glossed over. However, we require a more substantial and satisfying perspective.

Perhaps one core aspect of consciousness is *one's ability to perceive one's own existence as an independent coherent entity*. This is more satisfying, and perhaps, after a moment's reflection, the reader will accept it. If not, then ignore the rest of this section, as it will be of no use to you. (You don't exist!)

As you have reached this point you have hopefully (at least in part) accepted the above statement. However, on careful reading one glaring issue emerges. The thesis is self-referential—referring completely to itself. What are the implications here for the potential fabrication of conscious entities? Quite simply this—it is impossible for any conscious entity to prove conclusively the existence or non-existence of any other conscious entity. I cannot prove your existence as a conscious entity, the cat's existence, or indeed the existence (as *conscious* beings) of any of the other 7-billion-odd souls said to inhabit this planet.

6.1.1 Existence of Consciousness in Natural and Artificial Entities

Let's now look at some organisms that we may feel it plausible to assume have consciousness and see if we can determine some common traits or features that will enable us to come up with an even more concrete definition. The above perspective "the ability to perceive one's own existence as an independent coherent entity" may be reasonable from a philosophical and/or physiological viewpoint; however, scientists and engineers may wish to tie the matter down further. One possible route might be to guage the percentage belief we might have in the existence of conscious activity in different entities, both natural and artificial, in a manner analogous to the Turing test (Table 6.1).

So, using this general framework, we might place a threshold on the development of thought in a machine context by requiring the same percentage of probability of consciousness in that artificial entity as our percentage belief in the existence of consciousness in other humans (in this case 95%). Based on these tables we would

Table 6.1 Potential conscious activity—example tables showing the percentage belief one might have in the existence of consciousness for different entities, both natural and artificial

Natural	
Oneself	100%
Immediate family/friends	95%
Humanity in general	92%
Chimps, bonobos, orangutans, gorillas...	75%
The cat next door	30%
Beetles, spiders, etc.	2%
Plants	0.5%
Rocks, water...	0.01%
Etc.	...
Artificial	
IBM Watson	3%
AlphaZero	2.7%
AlphaGo Zero	2.5%
Deep Blue	2%
AlphaGo	2%
Etc.	...

conclude that our belief in the existence of conscious activity in even the most convincing artificial entities (in this example case IBM Watson) falls well short of our belief in the existence of conscious activity in humanity in general.

6.1.2 Delving Deeper

The previous discussion may have come across more as a statement as to what cannot be discovered about conscious entities rather than what can. Let us now discuss briefly what properties we would reasonably expect a conscious entity to possess in order for it to function as such. The notion of reality is a core concept, and one of the key issues in our discussion here. So, what exactly do we mean by reality?

Obviously, people's perceptions of reality vary. We would reasonably expect a blind person, for example, to have a quite different perception of reality from that of a sighted adult. A caterpillar would have a very different perception, or indeed a bat, as discussed in Thomas Nagel's highly cited article "What is it like to be a bat" (1974), (we have not excluded any object from the possibility of conscious existence at this stage). Let us then define reality for an individual or conscious entity (CE) A at a certain instant as "A's conscious perception of the world around him at that point in time". [We may, of course substitute "her" or "it" for "him" in this context].

What do we mean by the "world around him"? In general, we will take it that A's world is presented to him by the sum total of A's sensory input at that instant as perceived and interpreted by A's current state of mind.

Two points need clarification here; firstly, what do we mean by sensory input, and secondly how can we define A's "state of mind". By sensory input we mean all of the usual senses: sight hearing, touch, etc., and also including proprioceptive stimuli relating to body movement and/or position. Remember A has no way of proving the existence of any other conscious entities (B, C, etc.) and hence of other interpretations of reality. So external reality to A is indeed defined by the sum of its sensory input. Secondly, how do we define A's "state of mind? The simple answer is—we cannot. As we cannot even prove the existence of A as a separate conscious entity we can or certainly not propose to then define that consciousness or state of mind. However, looking to our own selves we may find it plausible that some of the factors facilitating our current conscious state are previous experiences (i.e. sensory inputs), also the time of day, drugs, etc.

6.1.3 The Conscious Perception of Reality

Let us recap. We have a conscious entity A in existence. For A to exist it must be part of a reality R. We can picture the situation as in Fig. 6.1:

In the diagram above R^A is depicted as A'a perception of reality R. Here A is placed inside reality R, as whatever is not "real" to A does not, in a sense, exist as far as A in concerned—except, of course, memories, etc. contained within A.

A then perceives R. As each CE's perception of reality will presumably differ to a greater or lesser extent, it is reasonable to assume that A, in perceiving R, forms an internal picture or model of R. As R may well be conceivably changing, so too A's internal model must continuously change.

However, something is missing from our current picture and that is that for A to exist it must form part of reality itself. Now A may not perceive itself as part of reality R and this is all very fine. Or is it? For if A does not perceive itself as part of reality R then to itself it does not exist and must then cease to function as a conscious entity in its own right. This may seem an extraordinary conclusion; however a moments reflection will show it to be true. [We note that other conscious entities—A cannot now exist as a separate conscious entity as we have shown—may clearly perceive A's physical existence, however this is not the point.]

Fig. 6.1 Conscious entity
A pictured as part of
Reality R

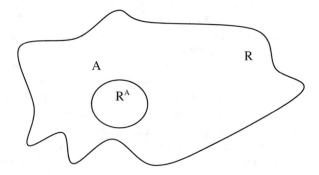

Fig. 6.2 Conscious entity
A pictured as part of R,
updated to include A's
conscious perception of its
existence as part of R

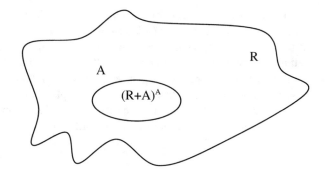

We must now replace our original figure in Fig. 6.1 with the updated version in Fig. 6.2 (assuming A to indeed have a conscious existence).

Here $(R+A)^A$ indicates A's conscious perception of its existence as part of R. However, in order for A to perceive itself as part of R it must have the ability change or modify R in some fashion. We can argue this case since if A is not possessed of this ability it then has no way of perceiving itself since R presents itself to A through its sensory input, and in this case (A not having the ability to modify R) A has no way of generating this input. A therefore ceases to function as a CE. So A must be able to both perceive and modify external reality R, and it is reasonable to conjecture that the more A can modify R, or the more power it has over its environment, then the more A in a sense <u>becomes</u> R, also the more "conscious" A becomes. Of course, the more ability A has to influence R the more A becomes R, until in the limit we have, in a sense, an omnipotent being with complete control over reality, and who is, in a sense reality itself.

6.2   The Formation of Understanding in the Rhythmic Nature of Reality

The rhythmic nature of reality may be viewed as one of the key factors in the formation of understanding and of consciousness itself. Reality as we perceive it is rhythmic in both a spatial and in a temporal sense. Some of these spatial and temporal rhythms may be imposed by the internal structures of the conscious entity (i.e. heartbeat, nature of the sensory organs, etc.), and some by the external environment the organism finds itself in, governed by the so-called "laws of nature", as observed and described in the physical and biological sciences. When we talk about rhythm here, the, the core point is *predictability*. The sciences only exist because of the predictable nature of reality. If one dropped an apple an couldn't predict what direction it would fall, or if the sun rose in the east some days and in the west other days, and so on, without a predictable pattern, then the sciences could not exist because they depend on the existence of repeating or repeatable phenomena.

In fact, in a universe of complete chaos or unpredictability it would be impossible for a conscious entity to exist (if indeed it would want to). For, as we have seen, one of the conditions for the existence of a conscious entity is the ability to manipulate and to modify its external environment. If this is manipulation in a true sense of its environment, then come consequences of this manipulation must, at some stage, be observable or repeatable. In a universe of complete unpredictability and chaos this could not be. Also, in a certain sense therefore the development of conscious entities might be viewed as a *consequence* of the ordered environment they inhabit.

Chapter 7
Computers and Thought

Strong AI holds that a computing machine with the appropriate functional organization (e.g. a stored-program computer with the appropriate program) has a mind that perceives, thinks, and intends like a human mind Philosophy of Science –An Encyclopaedia—*on the connection between computation and thinking (Sarkar and Pfeifer 2006).*

7.1 Thinking Versus Calculation

In his preliminary remarks to lecturers, Dr. A. M. Uttley (N.P.L.) suggested that the hoary question 'Do machines think?' should be avoided; and the feeling of the gathering was that the remark of one delegate 'No, they are like women, they calculate' was as good a reply to any to such an ill-defined question.

So began the rather caustic (some might even say misogynistic) introduction by Dr. Uttley to the influential Symposium on the Mechanisation of Thought Processes (Blake and Uttley 1959), held in England in November 1958. (The N.P.L. in the above quote refers to the U.K. National Physical Laboratory.) To the modern perspective this viewpoint might indeed appear a tad misogynistic, but we should perhaps approach this from the perspective of the late 1950s, when the vast majority of researchers and innovators in the field of computation and the (newly christened) AI field were male. This is a particularly interesting perspective as it was, in fact, mainly females that were employed to operate the early mechanical calculators in the 1930s and 1940s. However, the history of people as computers goes back well before this to the seventeenth century and the numerical calculation of the orbital trajectory of Halley's comet (Grier 2001). Also, in many cases these pioneering human computers were men or boys.

But could a machine actually *think* rather than merely *calculate*? That was the question. On the face of it the two operations require many similar facilities. Both have a memory requirement, in order to store the results of past musings (or calculations), both require certain logical faculties. However, and crucially, thinking also requires the ability to take decisions based on past information, and also the ability to *reason*; that is to consider and potentially resolve arguments using

© Springer Nature Switzerland AG 2020
M. Eaton, *Computers, People, and Thought*,
https://doi.org/10.1007/978-3-030-55300-5_7

logical and rational means—abilities certainly not generally associated with mere calculating devices.

7.1.1 Computers, They Just Do What They're Told: Don't They?

We may argue that computers only do what we tell (program) them to do—surely this cannot be viewed as *true* intelligence.

Well—yes and no. From the earliest days of AI (before the term was even coined), researchers were seeking to implement learning capabilities on a computational substrate. One of the earliest examples of this was Samuel's 1959 checkers-playing program, still impressive today, which learned over time to adapt its play based on experience. True, the application was "programmed", however in a sense, over time, it learned to "rewire" its programming (Samuel 1959). In a similar fashion (although not completely analogous) modern so-called deep neural networks learn to change and adapt their structure in response to training inputs, allowing for, for example, better than human-level object recognition in some instances.

But even without learning capabilities computers can exhibit characteristics that most of us would describe as intelligent behaviour in a broad sense. Deep Blue, the chess playing computer that defeated the world chess champion Garry Kasparov in 1997 had no learning capabilities, it was "just" a highly sophisticated chess-playing machine.

7.1.2 Core Competencies Rather Than Intelligence?

A central notion/concept of this text is that of "intelligence". However, intelligence proves so difficult to define or to compartmentalise that perhaps it might be better to do away with it altogether? What exactly is it that distinguishes intelligent activity from simply a level of ability/competence in a particular area or topic (or group of topics)?

So if a major problem with the AI "field" is that if one can't even define its core concept— "Intelligence"—where does that leave us? Perhaps it is more helpful, in a general context, to deal with competencies/abilities rather than intelligence per se?

7.1.3 Outline Taxonomy of AIE Prowess

It is helpful to distinguish between a number of terms used to describe different levels of AI/robotic intelligence and competence. In Table 7.1 we give a brief

Table 7.1 Explanation of some commonly used terms in this text

AIE: Artificially Intelligent Entity	An artificially intelligent agent, competent in a particular field, or number of fields. May be embodied (a robot of some form)
A^2IE: Advanced Artificially Intelligent Entity	An agent, possibly embodied, operating at human-level capability (or possibly beyond) in many (or all) areas of human competence.
A^3IE	Superintelligence. General intelligence level well beyond human level
AGI: Artificial General Intelligence	A^2IE level intelligent agent
HLAI: Human-level Artificial Intelligence	A^2IE level intelligent agent
BFH: Built for Human	An embodied AIE able to operate in most environments designed for human inhabitants, but with limited capabilities
HI: Human-inspired	An embodied AIE with a morphology broadly similar to that of humans
IH: Inferior Humanoid	An embodied AIE with the broad morphology of humans—two arms and legs, stereo vision and auditory sensors, etc.
Humanoid	An embodied AIE/A^2IE with high levels of intelligence and dexterity and a morphology similar to that of humans
Android	An embodied A^2IE with appearance on behaviour very close to that of humans
Replicant	An embodied A^2IE with appearance and behaviour identical to that of a human
Superintelligence	An A^3IE agent
Superhuman	An embodied AI^3E in human form
Artilect: Artificial Intellect	[after de Garis]. An A^3IE level intelligent agent

summary of a number of these terms, some in common usage and some which have been introduced in this text. Our taxonomy is broadly based on level of competence and/or anthropomorphicity (humanlike qualities). Some of these terms were originally introduced in Eaton (2015).

7.1.4 What Is (and What Is Not) AI

While introducing the topic of AI to third-year undergraduate students I have been struck by the difficulty of defining or delineating the area: indeed, the bones of one or two lectures could be spent on this very topic. Of course, part of this difficulty comes from the definition of the term intelligence itself. What exactly does it mean for an entity (either "Artificial" or "Natural") to be in possession of this quality?

Several "definitions" of the AI/IS field are currently extant. One of particular interest is that AI deals with the aspects of intelligent activity at which, at the moment, humans are better than machines. This poses several interesting questions, not least of which is—what exactly *is* it at which currently humans are better? We

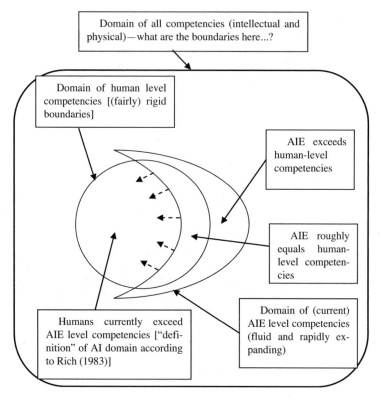

Fig. 7.1 Human versus Artificially Intelligent Entity (AIE) competencies

might pose this as a task—to take some area in which at the current time humans are better, and see what the current state of the art in AI in this area is, how close it is to human ability, and, given the current state of progress, when it is estimated that computers/robots will reach and/or exceed human ability in this particular field.

By any measure, the range of tasks for which humans equal or exceed machine capabilities is shrinking rapidly (see Fig. 7.1). This figure illustrates the gradual (and in some cases not so gradual) erosion of areas in which humans currently exceed AIE-level competencies.

7.1.5 AI Has Delivered?

Many (generally uninformed) commentators on the field claim that AI has not delivered many concrete results over the years. AI *has* delivered. It's just that in the past the delivery has not been as dramatic or as immediate as predicted by some commentators. However, as a "founder" of the AI field John McCarthy put it (as quoted in (Harvey 2013)) "As soon as it works no one calls it AI any more".

We might now transpose this to "as soon as it accrues human-level/human-superior performance/competence no one calls it AI anymore".

For example, why should research into a particular game domain generally either cease or be reduced considerably when a computer's ability in that game reaches or exceeds that of the best human? Given that in virtually all areas of intellectual pursuit it is highly likely that machines will outstrip humans in not too many years, the only real end point here is when the game has actually been solved, that is, for example for a board game, the provably best move can be generated for every possible board configuration. Of course, some games involving cyber-physical play, such as (embodied) soccer may never be actually "solved" in this particular sense.

Even today it is an interesting question—in 3 s name a particular task at which humans are currently better than machines? We can (generally) give a correct answer within this timeframe, but it may not be immediately obvious. I think that it is probably fair enough to say that within the next decade or two any disembodied game, or endeavour that can be so precisely defined that we may construct a clear unambiguous algorithm for it, has the potential to be tackled at least as well by artificial technologies as by "unassisted" human means.

7.1.6 Machines Need Not Apply

However, there *is* one thing that machines will never *ever* be better than humans at. What can this be?

Simply—*being human.*

No matter how we create mechanical devices to closely emulate or even to surpass humans, in both an intellectual and a physical sense, up to and including replicant level, a machine will never be more human than a human. And there lies the crux—in some sphere's humans will *always* prefer to deal with other humans. This being said, it is undoubtedly true that sometime in the (probably not so distant) future there may well be agents that appear more "human-like" than the "average" human. There is an interesting discussion on this general topic in Brian Christian's book *The Most Human Human* (2011).

But—however much a machine may look like or behave like a human; it will never actually *be* a human. And, for many of us, hopefully, this will remain a critical factor. I use the term hopefully advisedly, because in some circumstances it appears that people actually prefer dealing with a machine rather than a human. For example recent research indicates that business travellers prefer to use digital services rather than interact with human operators for hotel reservations (78%), for booking flights (69%), and for checking-in for flights (68%).[1] Another research study, based on Irish shoppers' preferences involved a nationally representative sample of 1070 shoppers,

---

[1] https://news.carlsonwagonlit.com/pressreleases/in-human-vs-machine-cwt-study-finds-two-thirds-of-travelers-prefer-machines-when-booking-air-travel-2876426

where 83% of 18–35 year olds indicated a preference for self-service checkouts at supermarkets as opposed to cashier checkouts. Overall 60% of shoppers indicated a preference for self-service checkouts. These figures, and others like them, do not appear to bode well for human participation in the workforce over machines in many employment categories.[2]

7.1.7 Problem: What Problem?

AI is typically characterised as an approach or set of approaches to solving a problem or suite of problems, probably (though not necessarily) interconnected, or to achieving a (set of) goal(s). But what if there *is* no problem?

Man, himself is in a sense our touchstone and our proof of existence of intelligence (albeit in the natural world), and he does not come with a pre-written problem or set of problems to solve. Survival, certainly, propagation of the species, perhaps—but can these really be characterised as "problems" in the ordinary sense of the word?

Rod Brooks, the eminent roboticist and AI researcher and former head of the Massachusetts Institute of Technology (MIT) CSAIL (Computer Science and Artificial Intelligence Laboratory) put it as follows in his influential 1991 article "Intelligence without representation" (Brooks 1991), where he talks about the development of artificial intelligent autonomous entities he calls *Creatures*, which are created without any specific goal or set of goals, other than to exist in harmony with humans:

> I wish to build completely autonomous mobile agents that co-exist in the world with humans, and are seen by those humans as intelligent beings in their own right. I will call such agents Creatures. This is my intellectual motivation. I have no particular interest in demonstrating how human beings work, although humans, like other animals, are interesting objects of study in this endeavour as they are successful autonomous agents. I have no particular interest in applications; it seems clear to me that if my goals can be met then the range of applications for such Creatures will be limited only by our (or their) imagination. I have no particular interest in the philosophical implications of Creatures, although clearly there will be significant implications.

In their recent, widely acclaimed, textbook *Why Greatness Cannot be Planned* (Stanley and Lehman 2015), Kenneth Stanley and Joel Lehman argue that if we wish to address difficult and complex areas of endeavour, such as, for example, the achievement of human-level intelligence in a machine, it may in fact be counterproductive to set detailed and specific objectives to be achieved, but rather preferable to seek out behaviours that are novel and unseen. They further extend their discussion to the societal level; arguing that the increased drive to quantify and measure so much of what we, as humans do, serves to both dehumanise society and to stifle creativity. These are areas that we will return to later in this book.

---

[2]https://www.checkout.ie/retail-intelligence/six-10-shoppers-prefer-self-service-checkouts-cashier-checkouts-54113

7.2 AI as Representation and Search

Classical approaches to Artificial Intelligence have long been dominated by the notion of AI as having *representation* and *search* at its core. Based on this approach we first address the problem domain in question, be it in the field of robotics, natural language understanding, or whatever, and abstract from this problem domain the core attributes required for its representation in the computational substrate being used in order to address the particular problem domain. A key issue in this abstraction is to select just those components, and no more, required to effectively and efficiently represent the problem domain in question. For example, in the case of a chess-playing robot we are generally not at all interested in the exact shape of the individual pieces (as long as we can distinguish rook from bishop, and so on), however it *is* essential that we are aware of the different movement patterns of these two pieces, and of all of the other chess pieces on the board, and, of course, also of the overall rules of the game of chess being played (be it western-style, Japanese, Chinese, etc.)

Once we have represented the problem domain within the artificially intelligent entity (AIE), we can then perform a search through the sequence of possible combinations, starting from the initial state, in order to reach a goal state representing a successful outcome for our AIE. This process is known as *state-space-search*. Again, taking the game of chess as an example, the start state is given, with the pieces of each opposing player on opposite sides of the board in a particular configuration—rooks at the corners, king and queen at the centre, pawns to the front, etc. Eventually, after a sequence of moves, a goal state is reached once we checkmate our opponent. We will now briefly look at each of these two components, representation and search, in some more detail.

7.2.1 Representation

The essence of a good representational scheme is to abstract out those components of the problem domain (and only those components) that are necessary for the problem-solving mechanism to be effectively applied. Here we have a trade-off between *efficiency* (how well these components are extracted without extraneous unnecessary material), and *expressiveness* (how closely the abstraction relates to the actual features in the problem domain) (Luger 2009, p.37).

As a simple example let us say the constant 2.7181828 crops up in the problem domain. We might choose to represent this value in its entirety (in binary format, of course, for a binary computer). Here we trade off efficiency for expressiveness. For the problem in hand we might only require to represent a crude approximation to this value, let us say to two decimal places, or 2.72. Here we have an efficient representation which, we judge, sufficient to our needs. Of course, if our intention is to represent the transcendental number e, in this case full expressiveness can never be achieved as we cannot encode the exact number given its transcendental nature. In a similar fashion we could choose to represent a pictorial image such as that presented in Fig. 2.1 in Chap. 2 using more or fewer pixels, depending on the requirements of our problem-solving mechanism. For example, if our task is an image recognition one to recognise a horse from a cow, it is likely that we will need to use a much higher resolution image than, for example if our task was to distinguish a handwritten letter "a" from a handwritten "z".

7.2.1.1 A River-Crossing Puzzle

As a simple example of the use of an appropriate representational mechanism let us look at the so-called Farmer, Fox, Goose, and Grain puzzle. This puzzle has ancient origins in both the West and also in African culture. The origin of the Western version (sometimes substituting a Wolf for the Fox, a Goat for the Goose, and a Cabbage for the Grain) is generally attributed to the intellectual Alcuin of York (735-804), and contained in his treatise *Propositiones ad acuendos juvenos* or, in translation, *Problems to sharpen the young*, which consisted of a set of 53 numbered problems, with answers to the problems also supplied.

The puzzle, as originally presented, is quite straightforward. We have four actors (here we will use the term "actor" as proposed by Russell and Norvig (Russell and Norvig 2010, p.426) as "a participant in an action or process"), which we wish to transport from the left to the right-hand side of a fast-flowing river using a rowing boat in the most efficient manner; i.e. the smallest number of river crossings. The construction of the boat is such that a maximum of two of the actors can be in the boat at any one time, and there are the extra constraints that the fox cannot be left alone with the goose (for obvious reasons—the fox will devour the goose) or the goose left with the grain (a similar scenario), and (of course) that the boat can only move with the farmer in it.

Our task now is to transport all four safely across the river in the shortest timeframe. We now seek a compact and efficient method of representing the core issues in this simple problem, just focusing on the important aspects and discarding all unnecessary details. I use this example as an introduction to the area of representation in AI with my third year "Intelligent Systems" students; the initial question that I ask is "What elements in the problem domain may be safely ignored or discarded?" After a short period of reflection it is clear to most students that we can safely do away with both the river and the boat in our problem representation; it is sufficient to simply represent the current positions of each of the four actors, on either one side of some divide or the other. Invariably, also, some student is quite clear that the farmer can also be excluded from the system as "he always has to be in the boat". On closer inspection, however, this proves not to be the case, as the position of the farmer indicates the current side of the river the boat is on and tells us the next direction of movement across the river. And, of course, crucially, the presence of the farmer prevents any unwanted banqueting on whichever side of the river she or he is on.

What is, perhaps, not so obvious on initial inspection is that is only required to represent the actors on one side or the other of the river bank, as we can assume that any not on one side will have made it safely across to the other (assuming no unfortunate boating accidents).

7.2.1.2   Representational Scheme

In our exposition we will choose to represent just the actors on the left-hand side. So, for example, using a single letter of the alphabet to represent each of the actors the state with the Farmer and Goose on the left-hand bank, with the foX and the graiN on the other can be represented as [F,G]. Of course, we could also use binary numbers to represent the presence or absence of each of the actors, but this might tend to obfuscate a little our current analysis. So clearly now we can dispense with both the river and the boat—assuming, of course, a safe passage is assured in both directions.

7.2.1.3   Complexity of the Search Space

In order, now, to formulate this problem in a manner amenable to our search mechanism, it will be helpful to identify the *valid states* in our problem domain. The total *state-space size* gives us an idea of the overall complexity of our problem domain by enumerating the total possible valid states. For example, a recent estimate (Steinerberger 2015) is that there are around 10^{40} different possible valid board states in chess.

Another method of estimating the complexity of our problem domain is to estimate the total number of games playable, also sometimes known as the *game-tree complexity*. For example, in the game of chess White has a total of 20 possible opening moves (two possible moves for each pawn, and two moves for each of his knights), and Black has 20 possible moves in response, giving a total of 400 possible board positions after just a single move by each player. (In general, a move by either player is described in the literature as a *ply*; some researchers use the term to denote a move by a player together with their opponent's response; however, we will use the first definition—one move by either player.) So, it is clear that this number grows very rapidly indeed for the game of chess—it is estimated to be on the order of 10^{120}, more than the number of atoms in the universe! This number is also sometimes known as the *Shannon number*, after Claude Shannon whose ground-breaking 1950 paper "Programming a computer for playing chess" (Shannon 1950) paved the way for the computing and AI developments that would eventually lead to the defeat of the reigning world chess champion Garry Kasparov in 1997 at the hands of the Deep Blue chess machine (Campbell et al. 2002).

Getting back to our earlier example, we can deduce, on initial analysis, that an upper bound on the state space for the farmer, fox, goose, and grain problem is 2^4 or 16. This is clear from our representational mechanism, where each of the actors may be present or not on the left bank of the river. On further analysis 10 of these 16 states are legal; the other 6 are excluded because of constraints imposed: e.g., the fox cannot be left alone with the goose without the farmer present, etc.

7.2.1.4 Legal Operators and the State-Space Graph

Based on these 10 valid states we now want to come up with legal operators that will move us through the state space, from the initial state (all four actors on the left bank of the river) to the goal state (all on the right), in the shortest possible time. Based again on the constraints imposed we arrive at eight legal operators, with the farmer, either on his own or with one of the other actors, travelling from either the left bank to the right, or in the opposite direction. We can quickly whittle these eight operators down to four (Table 7.2), based on the observation that the direction of travel is dictated by the position of the farmer—if the farmer is on the right bank, only right-to-left travel is possible, and vice versa.

Of course, not all of these four operators will be applicable in all states. For example, in the state [F, G, N] operator 4 will not be applicable, as the farmer and the fox are on opposite riverbanks. This is analogous to the game of chess, where, for example, the rook is legally allowed to move up and down or left and right on the chessboard as far as the player wishes—but only if there is not a piece blocking its way. These operators then form the links between the individual nodes of the graph.

In fact, chess is not the best example in this regard because of the presence of an unpredictable adversary, but we will shortly look in a little more detail at a non-trivial illustrative example without this complicating adversarial component— the 15-puzzle. We will shortly address the topic of search mechanisms specifically adapted for adversarial game play in the section on game AI. Now that we have defined the state space and legal operators it is a relatively straightforward exercise to draw the final diagram. This structure, obtained by combining the nodes in the graph with the links representing valid moves between individual states, is known as the *state-space graph* for our problem domain (Fig. 7.2).

Table 7.2 Valid operators for Farmer, Fox, Goose, and Grain puzzle

Operators:
1. F ⟺
2. F, G ⟺
3. F, N ⟺
4. F, X ⟺

Fig. 7.2 The complete
state-space graph for
Farmer, Fox, Goose, and
Grain puzzle

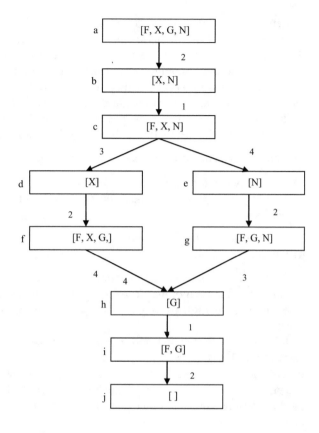

7.3   Search

Once we have settled on an appropriate representational mechanism for our problem,
the next (final) step is to perform a search through our state space in order to reach
the goal state from the start state, ideally with the minimum computational effort. Of
course, there may be a number of possible goal states (as in chess where there are a
huge number of possible winning states)—our aim is to reach one of those winning
positions with the minimum of effort. Generally, the lower the number of states
examined, the lower the computational cost, although, as we will see, this may not
always transpire to be the case. Another important measure of the effectiveness of
our search algorithm will be its ability to guarantee that, once a path has been found
to the goal state, this is the *shortest* path, in terms of the number of nodes that will
need to be traversed from the start node in order to reach the goal node on any
subsequent traversal of the graph. Any algorithm that guarantees to find this shortest
path is known as an *admissible* algorithm (Hart et al. 1968).

Another, more stringent criterion would be that of *optimality*, by which we mean
that no other admissible algorithm examines fewer states in the state space.

Let us now briefly outline the basic procedure for state-space search. In our discussion we will use the term *state* and *node* interchangeably. We will assume that as the search progresses, we maintain two sets, the *open* set contains those nodes we currently wish to examine to see whether they are the goal state, and if not to *expand*. By expanding a node, we may generate one or more successor (child) nodes, which we add to our open set for examination. Finally, we will add the node which we have just examined to a separate *closed* set to indicate that it has been examined (and found not to be the goal). The basic algorithm proceeds as follows:

1) Choose a state from the *open* set as the current state (or if at start, use *start* node).

2) Check whether the current state is the goal; if so end search. If not, expand it by applying any appropriate and valid operators, and adding child nodes to the *open* set.

3) Add the current state to the *closed* set.
4) Go to 1.

We will also wish to take account of situations where we might revisit nodes that are already in the open or the closed sets, and also to maintain a record of the path taken, in order to allow us to reach the goal node as quickly as possible on future occasions.

In many cases the re-application of the same operator will return us to the state we came from. This is the case in the farmer, fox, goose, and grain puzzle. Because of the simplicity of the search space we can easily graph the complete state space for this problem (see Fig. 7.2); this will not however be desirable, or indeed possible, for most interesting problems, which generally will be of a considerably more complex nature.

As we can see, in a sense, once we have arrived at an efficient representation for this simple problem it almost solves itself, demonstrating the power of an efficient representational scheme. Starting with the initial state [F, X, G, N], where all of the actors languish on the left bank, only one of the four operators (all of which are applicable in this state) lead us to a valid subsequent state. Operators 1, 3, and 4 lead to states where either the fox is left alone with the goose, the goose with the grain, or all three together. We have already identified these as being outside our 10 identified legal states. Moving on to state [X, N], the re-application of operator 2 just brings us back to the original state, and operators 3 and 4 are not applicable in this case; again, this leads to only one possible subsequent state. In fact, the only choice to be made is in state [F, X, N], which has two branches, either of which leads directly to the goal state in a further four moves.

7.3.1 Informed and Uninformed Search Strategies

Using the basic search algorithm presented above, the main issue now to be decided in the implementation of this algorithm is in the *choice* of which node to expand next (assuming there is more than a single node left in the open list). We will briefly look now at several possibilities.

7.3.1.1 Random Choice

The first of these is *random choice*. This may seem, on the face of it, not a very promising selection mechanism; however, with certain constraints and enhancements this can, in fact, prove a surprisingly effective mechanism. Take a game of chess (or draughts) for example. White is in a certain position on the board and is to play next from a choice of, say, 15 possible moves. One approach would be to assume White makes one of these moves, and then play out the game assuming *random* moves from both Black and White until a final result is reached. We now repeat this process a large number of times, finally averaging the overall results to obtain the overall ratio of wins to losses (or draws) based on playing this particular move. Now follow the same process for another of White's possible moves, noting again the overall outcome, and repeat this procedure for all 15 possible moves. Finally play the move that has resulted in the highest winning ratio for White. This so-called *Pure Monte Carlo* approach, while on the face of it not appearing particularly effective, in fact forms the basis for many successful modern approaches to two-person adversarial game play.

7.3.1.2 Depth-First Search and Breadth-First Search

If we do decide to adopt a more structured approach, two well-known mechanisms come into play. These are *depth-first search* (DFS) and *breadth-first search* (BFS). We will examine each of these search mechanisms briefly.

Essentially depth-first search involves getting deep into the search space quickly (as its name implies), while breadth-first search adopts a more layered approach, exploring all nodes at a particular level in the tree before moving on down to the next level.

For our example earlier, and assuming nodes are opened in a "left-to-right" fashion, the succession of open and closed states for depth-first search would look thus:

1. open =[a]
   closed = []
2. open=[b]
   closed=[a]
3. open=[c]
   closed= [a, b]
4. open= [d, e]
   closed= [a, b, c]

5. open= [f, e]
   closed= [a, b, c, d]
6. open= [h, e]
   closed= [a, b, c, d, f]
7. open= [i, e]
   closed= [a, b, c, d, f]
8. open= [j, e]
   closed= [a, b, c, d, f, i]
9. success, j= goal!

The breadth-first trace, by contrast, would look as follows:

1. open =[a]
   closed = []
2. open=[b]
   closed=[a]
3. open=[c]
   closed= [a, b]
4. open= [d, e]
   closed= [a, b, c]
5. open= [e, f]
   closed= [a, b, c, d]
6. open= [f, g]
   closed= [a, b, c, d, e]
7. open= [g, h]
   closed= [a, b, c, d, e, f]
8. open= [h] (as h is already in open)
   closed= [a, b, c, d, e, f, g]
9. open= [i]
   closed= [a, b, c, d, e, f, g, h]
10. open= [j]
    closed= [a, b, c, d, e, f, g, h, i]
11. success, j= goal!

It is clear that the breadth-first algorithm, in its layer-by-layer approach, requires two extra iterations of the search, while the depth-first approach goes directly to the goal. However, because of the layer-by-layer approach taken by breadth-first search in examining *all* nodes at previous layers before moving on to the next layer, it can be demonstrated that BFS is admissible—i.e. it is *guaranteed* to find the shortest path to the goal, assuming this path exists. This, however, is at the overhead of the requirement to store all nodes of the current layer, before moving on to the next. DFS, on the other hand, gets deep into the search space quickly, and may well find a goal state more quickly than BFS (as in this example); however, we cannot guarantee that this path, once found, will be the shortest one. A simple method to algorithmically implement BFS and DFS, is to maintain open and closed *lists* rather than sets. Now, when placing expanded child nodes on the open list always place these nodes on the right of the open list for BFS, and on the left for DFS, first discarding any child nodes that are already on the open or closed lists (assuming that we always now choose the leftmost node in the open list for expansion next).

Two other search algorithms worth mentioning briefly at this stage are *bounded depth-first search* (BDFS) and *iterative-deepening depth-first search* (IDDFS). BDFS seeks to overcome the problem that DFS may get caught-up deep in the search and works by simply putting a depth limit on the search. IDDFS performs a series of bounded depth-first searches with the bound depth starting at 1 and increasing at each iteration. Like BFS, IDDFS is guaranteed to find the shortest path to the goal, however in a depth-first fashion.

All of the algorithms we have looked at to date have one thing in common; whether choosing a node randomly, or using some predetermined picking order, none of these algorithms take any account of the possibility that we may have at our disposal some information relating to the particular problem under consideration as to the relative worth of individual states in terms of reaching our goal. In this sense they are *uninformed*. *Informed* search algorithms, on the other hand, do take account of such information, assuming this is available, and form the core of *heuristic search*, a topic of key importance to intelligent search strategies.

7.3.2   Heuristic Search

> Much of what we commonly call intelligence seems to reside in the heuristics used by humans to solve problems.
> —George F. Luger *Artificial Intelligence: Structures and Strategies for Complex Problem Solving*, p.21

We now look at informed search algorithms; that is, those algorithms that take account of information that may be available regarding the relative desirability of entering one particular state over another. For example, in a game of chess, while it might be considered more advantageous to capture a knight with a pawn, how much more advantageous would it be in terms of board position to use that same pawn to take your opponent's queen? The algorithms discussed in the previous section (DFS, BFS, BDFS, IDDFS) do not allow us to make any such distinction in the choice of which node to expand next from the open list. Informed search does, and, in doing so, in a sense, puts the *intelligence* back into our search, because we now need to decide what particular attributes one state may possess that gives it an advantage (or disadvantage) compared with any other. Of course, it is clear that one particular state, the goal state, has an infinite advantage over any other (assuming we arrive there at optimal cost); but how are we to decide on the relative advantage of one state over another in a more complex context? This is where heuristics come into play.

There are many different definitions of heuristics. A particularly cogent description was given in Slagle's article contained in the 1963 compendium *Computers and Thought* referred to earlier in this text, and also published in the *Journal of the ACM* that year (Slagle 1963):

...a heuristic method...is a method which helps in a problem's solution by making plausible
but fallible guesses as to what is the best thing to do next

7.3.2.1 A* Search

Many different approaches to heuristic search exist; we will focus on one particular
method, A* (pronounced *a-star*), as originally described in Hart et al. (1968), which
is a powerful and flexible search mechanism, and which has formed the basis for
many different alternative search methods. A* has the nice property that, given
certain (not very rigorous) conditions it is *guaranteed* to be admissible, that is to find
the shortest path to the goal, assuming that one exists.

A* relies on the availability, for each node n in the search space, of a heuristic
estimate $h(n)$ of the cost of getting from that node to the goal state. For A* to
function we also require to know the actual cost of getting from the start node to our
current node: $g(n)$. Once we have these two values this allows us to construct an
evaluation function, f(n), where

$$f(n) = g(n) + h(n)$$

This evaluation function, in actual fact, gives us an estimate of the cost of getting
from the start node to the goal node, while still passing through node n. The more
accurate our heuristic $h(n)$, the more accurate this estimate will be.

The derivation of this formula is quite clear-cut. What we are interested in is
the desirability of picking a particular state over another, which is simple if all of the
states we are choosing from are at the same level—this is given by the node with the
lowest $h(n)$ value. But what if we are comparing states at different levels in the
search tree? How then do we choose? Adding the $g(n)$ factor into our evaluation
function allows us to compare the relative worth of individual nodes that may have
similar heuristic values but are at different levels in the search tree.

Our search methodology is now quite straightforward. We use the same basic
algorithm as described in the previous section, however now, with each element in
the open list we also associate its $f(n)$ value, as calculated above, and choose for
expansion the node with the lowest value, that is, the lowest estimated cost. We can
do this either by sorting the open list at each iteration, in order of heuristic merit, or
by using a particular data structure called a *priority queue*, which will ensure that the
node is placed in the correct order in the open list. We then continue with the search
process, as before, until the goal is found.

7.3.2.2 The 15-Puzzle

Many readers will be familiar with the 15-puzzle, a sliding-piece puzzle involving
15 square tiles, generally numbered 1–15, that can slide around on a four-by-four
grid. The puzzle has its origins in the 1870s and for a period in the early 1880s

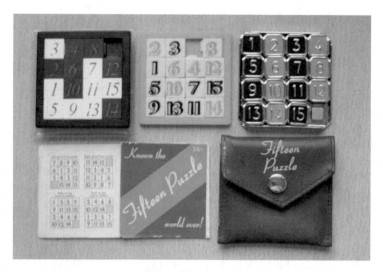

Fig. 7.3 A small selection from the author's extensive 15-puzzle collection

sparked a craze of interest, sparking newspaper headlines such as "15: The Diabolical Invention of Some Enemy of Mankind" (Slocum and Sonneveld 2006). It also spawned lyrics with such profound words as

I get number one in the corner, I get number two in its place,

When the number called three at the bottom I see,

And my steps I then have to retrace,

......

I've sat up all night but I've never got right,

That wonderful puzzle "fifteen".

The puzzle has also influenced some modern video games, such as the award-winning indie game "Cogs".[3]

The state-space size for the 15-puzzle may be calculated as 16!/2, as each of the 16 squares contains either one of the numbered tiles or a blank, and it turns out, on analysis that there are two non-overlapping sets of states; starting out in one of these sets it is impossible to enter a state in the other set without picking up a piece off the board and replacing it in a different position. This number enumerates as 10,461,394,944,000 reachable states in total, given a particular starting position; so, the 15-puzzle is not trivial by any means, and as such lends itself well to the application of heuristic techniques (Fig. 7.3).

---

[3]http://www.cogsgame.com/

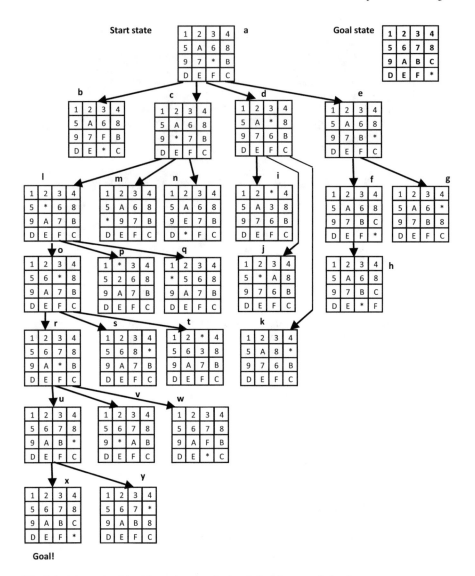

Fig. 7.4 A* search of the 15-puzzle using the heuristic "tiles out of place"

7.3.2.3 A* for Sliding-Piece Puzzles

In illustrating the operation of search algorithms as applied to sliding-piece puzzles most authors work with a pared-down version of the 15-puzzle—the 8-puzzle—but in our example here we will use the original full version. Puzzles like the 8- and 15-puzzle form useful testbeds for novel search algorithms and are also helpful as illustrative test examples in textbooks.

The following example (Fig. 7.4) illustrates the application of A* search to the 15-puzzle. The initial start state is shown at the top of the diagram. Successive open and closed states are shown below; the open list is sorted in each instance with the evaluated best state first in the list. For reasons of clarity and brevity we represent the numbers 10–15 by their hex equivalents A–F. The blank tile is represented by a *. The heuristic used is the "tiles out of place" heuristic, simply evaluated by counting the number of tiles in the position under consideration which are not in their correct place. So, for example, in the initial state, tiles A, 6, 7, B, and C are incorrectly positioned, giving an h value for this node of 5 and an f value also of 5 (as $g = 0$ for the initial state). Each of the stored values in the list give both the state name and the associated f value, so state b with an associated f value of 7 is listed as b7. The successive open and closed states for an A* search of the 15-puzzle are as follows:

1. open=[a5]
   closed= []
2. open= [e5, d6, c6, b7]
   closed=[a5]
3. open= [f5, d6, c6, b7, g7]
   closed= [a5, e5]
4. open= [d6, c6, b7, g7, h7]
   closed= [a5, e5, f5]
5. open= [c6, b7, g7, h7, j7, i8, k8]
   closed= [a5, e5, f5, d6]
6. open= [l6, b7, g7, h7, j7, i8, k8, m8, n8]
   closed= [a5, e5, f5, d6, c6]
7. open= [o6, b7, g7, h7, j7, i8, k8, m8, n8, p8, q8]
   closed= [a5, e5, f5, d6, c6, l6]
8. open= [r6, b7, g7, h7, j7, i8, k8, m8, n8, p8, q8, s8, t8]
   closed= [a5, e5, f5, d6, c6, l6, o6]
9. open= [u6, b7, g7, h7, j7, i8, k8, m8, n8, p8, q8, s8, t8, v8, w8]
   closed= [a5, e5, f5, d6, c6, l6, o6, r6]
10. open= [x6, b7, g7, h7, j7, i8, k8, m8, n8, p8, q8, s8, t8, v8, w8, y8]
    closed= [a5, e5, f5, d6, c6, l6, o6, r6, u6]
11. success, x= goal!

We mention a couple of interesting points about this example. Firstly, it is clear that we have arrived at the goal state, involving six moves from the start node: our path is a–c–l–o–r–u–x. Secondly, it should also be clear that we have found the optimal path using far fewer node evaluations than would be necessary for a breadth-first search—assuming, of course, we have indeed found the optimal path. DFS might have brought us to the goal with fewer evaluations; conversely it could have taken far more, depending on the order in which nodes were evaluated. And, of course, unlike BFS, DFS cannot guarantee that when we have found a path to the goal that this path will indeed be the shortest. So A* search does appear to have a definite advantage over both BFS and DFS, as we would expect, given the extra overheads involved in evaluating heuristics and in sorting the open list. A common error I notice in students studying the A* algorithm is that they assume that by following the estimated best path presented at each state the shortest path to the goal will automatically be found; i.e. there is no need to maintain open and closed lists. This is emphatically not the case. In our example the path a-e-f is followed, however when node h is reached with an f value of 7, the algorithm automatically backtracks to nodes c and d, which have a lower estimated cost, and the route to the goal is then found. It is clear from this example that the use of a heuristic does not mean that we will not end up going down "dead ends" before correcting our path to arrive at the optimal route.

However, one crucial factor is still missing. Although we have demonstrated that we need to examine far fewer nodes than for BFS (and, in general, this should be the case) we are still not certain that the path found is, in fact, the shortest one possible— a very desirable trait if we can demonstrate this to be true from our analysis.

7.3.2.4 Admissibility of A*

It turns out, in fact, that if A* is used in association with a heuristic with specific properties we can in fact guarantee the admissibility of A*. The condition is simply that the heuristic value generated for any node n, $h(n)$; is *always less than or equal* to the actual cost, $h*(n)$, of moving from node n to the goal state. In the 15-puzzle example above, it is clear that the tiles-out-of-place heuristic is always going to generate a value that is less than or equal to the actual number of moves required to move to the goal state (it will require at least one move, probably many more, to move each incorrectly placed tile into its correct position); hence our A* search is admissible, and the path that we have found *is* indeed the shortest one possible.

7.3.2.5 Outline Proof of the Admissibility of the A* Algorithm

As this is such an important result, we will provide here an outline "proof"; for the full mathematical version see Hart et al. (1968).

Remember $f(n) = g(n) + h(n)$, where n is any state in the search, $g(n)$ is the cost of reaching node n from the start state (in our examples here just based on the level of the tree), and $h(n)$ is the estimated cost, using our chosen heuristic, of reaching the goal state from state n. So, $f(n)$ is effectively the estimated cost of reaching the goal state from the initial state while passing through node n.

As we have stated above $h(n)$ is now guaranteed to be less than or equal to $h*(n)$, where $h*(n)$ is the actual cost of reaching the goal state from the present state.

Hence

$$f^*(n) = g(n) + h^*(n) \geq f(n) \text{ for all } n$$

Now assume we have reached the goal state g.

So, $h(n) = h(g) = h^*(g) = 0$ and hence $f(g) = f^*(g)$ (estimated cost = actual cost).

This value (by definition) is less than or equal to the value of any open node (because of the best-first search mechanism we have used to arrive at this state). However, all these values are optimistic (from (1) above) so any path passing through any of these open nodes will result in at least as high an actual cost; hence we must have found the optimal path. QED.

To further illustrate A* admissibility, let us look at the previous example again briefly. On arrival at the goal state, the open list looks thus:

$$\text{open} = [x6, b7, g7, h7, j7, i8, k8, m8, n8, p8, q8, s8, t8, v8, w8, y8]$$

Here x is the goal state, which becomes clear when we examine it on the next iteration. (Note that the goal state might present itself in the open list, however if it is not chosen for expansion (i.e. at the left of the list, assuming the list is sorted in best-first order) we have no guarantee that the shortest path has been found.)

However, it *is* chosen, and as such we are sure that all other nodes on the open list will have f values greater than or equal to it, which is clear by inspection of the open list. But we have now arrived at the goal, and any shorter path from the start node to the goal must, by necessity pass through one of these other nodes on the open list (b, g, h,. . .y). But, because we are using an admissible heuristic, we know that the *actual* cost associated with each of these nodes must be greater than or equal to the *estimated* cost, hence the optimal route has indeed been found.

7.3.2.6   Inapplicability of A* Search

We should note that A* and state-space search are not so applicable to problems where the search space is continuous or for function optimisation problems. In circumstances such as these alternative search mechanisms such as simulated annealing or genetic algorithms may prove more appropriate.

7.4   Data Mining

Data mining, as defined by Manyika et al. (2011) consists of

> A set of techniques to extract patterns from large datasets by combining methods from statistics and machine learning with database management.

Or, according to Witten et al. (2016)

> Data mining is the extraction of implicit, previously unknown, and potentially useful information from data.

The basic intent of data mining is clear from these brief descriptions. We have a body of data (representing numerical, pictorial, language, or some other type of information, as discussed in Sect. 7.1), which is potentially very big (however we define *big*), from which we wish to pick out patterns and form relationships.

There has been a not inconsiderable amount of hype associated with data mining and the associated term *big data*. Many researchers are unhappy with the use of the term "big data", considering it too amorphous or vague. After all, disk drives containing terabytes of data are commonplace nowadays, whereas only a decade ago this would have been considered a gigantic quantity of data. Perhaps a helpful perspective on the meaning of big data is that contained in the 2011 McKinsey report *Big data: The next frontier for innovation, competition, and productivity* (Manyika et al. 2011)

> "Big data" refers to datasets whose size is beyond the ability of typical database software tools to capture, store, manage, and analyze.
>
> ...
>
> We assume that, as technology advances over time, the size of datasets that qualify as big data will also increase. Also note that the definition can vary by sector, depending on what kinds of software tools are commonly available and what sizes of datasets are common in a particular industry.

The process of trying to discover patterns in data is not new—some of the early applications of Artificial Intelligence were in the field of pattern recognition—what *is* new is the sheer volume of data that is being produced in recent years. So, however we may deprecate the term "big data", by any stretch of the imagination the amount of data being produced worldwide these days is certainly big. If anything, the term understates considerably the volumes of data involved; nowadays the terms *huge data* or *gigantic data* might well be more appropriate.

To get some handle on the quantity of data being produced, according to Helbing et al. (2017), in 2016 alone as much data was produced as in all years combined before this, and by 2027 there will be as many as 150 billion separate measuring sensors, networked together and producing enormous quantities of raw data. Central to the process of many data mining applications are the twin areas of search and machine learning, both of which topics we deal with in some detail in this book. In fact, the subtitle of a popular data mining textbook *Data Mining* by Witten et al. (2016) is *Practical Machine Learning Tools and Techniques*.

A commonly used machine learning technique is C4.5 which is based on ID3 which we discuss in the section on symbolic machine learning. A typical data mining might use an off-the-shelf package such as the WEKA (Waikato Environment for Knowledge Analysis) workbench, available from http://www.cs.waikato.ac.nz/ml/weka, which constitutes a selection of machine learning algorithms and data pre-processing tools in a single package, taking much of the laborious work away from the data mining practitioner.

7.4.1   Overview of Data Mining Applications

We will discuss here briefly some typical current data mining applications. For further details on many of these application areas see Witten et al. (2016). Commonly used data mining applications include Web mining, which involves extracting interesting and relevant data from the World Wide Web (Singh and Singh 2010); text mining which involves looking for specific patterns in text which includes applications in natural language processing and document classification and clustering (Witten et al. 2016); image mining (Devasena et al. 2011) and video mining (Li et al. 2019), typically involving the recognition and classification of image and video data respectively, and the application of data mining techniques in adversarial situations, such as spam email, which is specifically designed to evade detection. Another common use of data mining today is in the area of recommender systems (Najafabadi et al. 2019), where based on a user's previously expressed preferences, they are presented with further choices that are considered appropriate to their interests in domains such as news articles, videos, and books.

These are just some of the applications of data mining currently in use, in the future there may be many others, and some consider that we may be headed to an era of so-called ubiquitous data mining where data collected from a wide variety of consumer devices is woven together bringing the digital world and the real world into ever closer contact.

7.4.2 Data Mining: Ethical Issues

Clearly with data mining technologies with the potential described, problems do exist. For example, many people provide information online with the expectation that they cannot be identified, given the small amount of personal data they provide. However, it transpires that over half of all Americans can be identified given just their city of residence, their birth date, and their sex (Witten et al. 2016).

Another area fraught with ethical implications is that of image (and video) recognition particularly in the area of facial recognition technologies (Wang and Deng 2018). Using a technique called deep convolutional neural networks, better than human level performance has been achieved. While this may be seen as a boon for security and law-enforcement agencies, as well as for governments which may, for whatever reason, wish to monitor and track the movements of their citizens, the implications for privacy of the ordinary citizen are profound together with, of course the potential for the misuse of such technologies in the hands of criminals. We will discuss these important issues further in Chap. 13. As Witten et al. (2016) concisely put it

> ...the techniques described ...may be called upon to help make some of the most profound and intimate decisions that life presents. Data mining is a technology that we need to take seriously.

7.5 Game AI

> The AI must be entertaining, and to achieve this aim, it must more often than not be designed to be suboptimal. To be enjoyable, an AI must put up a good fight but lose more often than win. ...

I am firmly of the opinion that if the player believes the agent he's playing against is intelligent, then it is intelligent. It's that simple. Our goal is to provide the illusion of intelligence, nothing more.
—Buckland (2005)

7.5.1   What Is Game AI?

Anything that gives the illusion of intelligence to an appropriate level, thus making the game more immersive, challenging, and, most importantly, fun, can be considered game AI.
—Bourg and Seeman (2004)

There are several views as to what exactly defines and delineates the general field of game AI. The above two descriptions summarise what many people see as the core of Game AI today. But, of course, game AI also encompasses the tools and techniques used by Deep Blue against Garry Kasparov in 1997, and Samuel's ground-breaking 1959 learning checkers-playing program. Many also consider that areas such as procedural content generation (PCG) where, for example, entire game maps and levels, or portions thereof, are generated automatically, fall under the general remit of Game AI.

In using AI to reproduce/recreate human-level capabilities we should always remember that human-level capability is not optimal in most spheres. So, creating believable human-like abilities is, in most cases, not the same as creating the optimal opponent or strategy, or, indeed, physical body/mechanical structure.

This is often a major factor in Game AI design, where often the focus is on the creation of *believable* human-like (or animal-like) behaviours, rather than "optimal" behaviours, which, more often than not, may reduce (in some cases significantly) the entertainment value of the game.

The topic of game AI is itself the subject of several books—we refer the interested reader to the excellent *Artificial Intelligence for Games* by Millington and Funge (2009) together with the *AI Game Programming Wisdom* series, edited by Steve Rabin (2002–2014).

In this section we will focus on two important aspects of game AI. The first is the issue of *pathfinding*, which is of crucial importance to many modern game genres, allowing AI characters to generate efficient and believable paths through 2D and 3D game environments in search of goals such as food, weapons, or even you. The second area we will address is that of two-person adversarial search—this is the type of search required for playing games such as chess, draughts, and Go. Also, in a little while, in the section addressing the important topic of learning we will look at an example of the reinforcement learning paradigm applied to the game of tic-tac-toe (noughts and crosses).

7.5.2   Pathfinding Using A*

One of the most important applications of artificial intelligence in modern computer games is that of pathfinding. Pathfinding also has many obvious applications in the robotics field. Pathfinding involves the generation of sensible and possibly (though not necessarily) optimal routes for game characters in order to get from their current position to some goal position. The goal may consist of food or a weapon, or perhaps some exit point in the game level (Millington and Funge 2009). En route to the goal the AI character's path may be blocked by various obstacles, some impenetrable and some perhaps not.

Here will briefly look at navigation on a grid-based system, moving from tile to tile, where our search space is divided into a regular grid of small square tiles. These tiles may also be triangular or hexagonal.

The A* algorithm, discussed earlier forms the basis for the vast majority of pathfinding algorithms in common use today. In order to apply A* to find an optimal path we must use it with an associated admissible heuristic. One commonly used heuristic uses the *Manhattan distance metric*, so named after the grid-shaped street structure of central Manhattan. We calculate this simply as

$$h(x, y) = \mid x - x\_Goal \mid + \mid y - y\_Goal \mid$$

where (x, y) is the position of the current tile under consideration and (x_Goal, y_Goal) is the position of the goal state. It is clear that this heuristic will always generate a value that is less than or equal to the distance between our current position and the goal; hence this is an admissible heuristic and guaranteed to find the optimal path. Other possible admissible heuristics include the *diagonal distance metric*, if diagonal movement is allowed, and the *Euclidean distance metric*. Both of these heuristics are admissible but come at a higher computational cost.

We will use an example that involves three different types of obstacle, an *impenetrable forest*, which, as its name says, is impassable and cannot be entered; a *lake*, which can be traversed, however at twice the cost associated with normal terrain; and some *hills*, which, again, can be traversed, but at five times the cost. It is important that, as we work through this example, we use the actual cost incurred for the g(n) function (that is, including any additional costs that may be incurred by moving through the lake, or the hills), whereas for the h(n) estimate we ignore these potential costs (as, indeed we would also ignore any impenetrable obstacle). (Note: to avoid any potential confusion or ambiguity, we will say that we incur any additional costs associated with a particular state when *entering* that state from a left-to-right, or a top-to-bottom direction, and when *exiting* that state, if going in a right-to-left or a bottom-to top-direction.) See Fig. 7.5 for the map used in this pathfinding problem.

The succession of open and closed states for this problem will look as follows:

1. open[start2]
   closed []

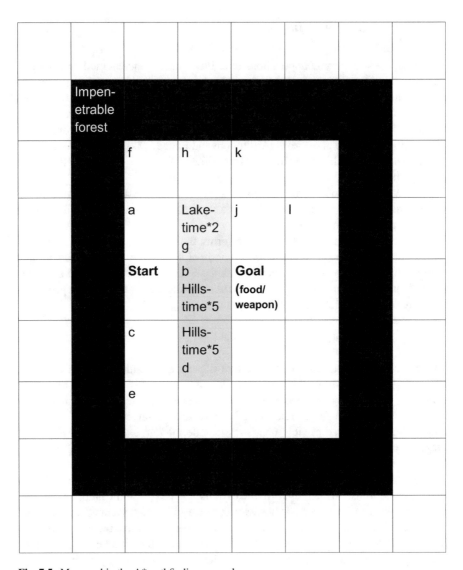

Fig. 7.5   Map used in the A* pathfinding example

2. open [c4, a4, b6]
   closed[start2]
3. open [a4, b6, e6, d8]
   closed [start2, c4]
4. open [g5, f6, b6, e6, d8]
   closed [start2, c4, a4]
5. open [j5, f6, b6, e6, h7, d8]
   closed [start2, c4, a4, g5]

6. open [goal5, f6, b6, e6, h7, k7, l7]
   closed [start2, c4, a4, g5, j5]
7. success, goal reached!

The final computed path is start2-a4-g5-j5-goal5, and on inspection this is clearly the path with the lowest cost—for example, going directly to the right and over the hills will incur a cost of 5+1=6. Similarly avoiding the hills and the lake altogether by going up or down, and around, will incur a total cost of 6.

Our basic approach may be extended to terrain with many different types of obstacles, either in 2D or 3D. We can also apply this approach to triangular-or hexagonal-tiled areas, and use polygonal representations, or structures termed navigation meshes, which encapsulate the navigable areas of our map. For further discussion of these various possibilities see the excellent *Artificial Intelligence for Games* book by Millington and Funge (2009).

7.5.3   Variants of A*

As mentioned earlier, because of its power and flexibility, and particularly because we can prove the admissibility of A* under the conditions outlined, A* has spawned many variants, which we do not have the space to discuss in detail here. Most research in the pathfinding field in recent years has been concerned with the development and the refinement of such algorithms. Two such variants are *beam search*, which is essentially A* with a limited open list, and *Dynamic A* or D* (Stenz 1994), which allows for pathfinding in partially known or changing environments (standard A* requires the provision of a completely specified world map *a priori*).

Another commonly used variant of A* is *Iterative Deepening A* (IDA*)* (Korf 1985), which itself has been subject to refinements and enhancements in recent years. IDA* has the advantage over A* that it is not necessary to maintain a sorted list or equivalent ordering mechanism; also, because it essentially involves a series of depth-first searches, memory requirements are lower than for A*. However, like A*, IDA* is guaranteed to result in an optimal solution, if this exists.

The basic IDA* algorithm operates as follows:

- Perform a series of depth-first search iterations starting from the initial state
- Explore each path until $f(n) = g(n) + h(n) >$ iteration threshold
- The iteration threshold for each level $i+1$ is set as the minimum cost of all of the nodes on the frontier of the previous generation i
- The threshold for the first iteration is just the heuristic estimate, $h(n)$, for the initial state
- The algorithm concludes when the goal node is chosen for expansion

7.5.4   Application of IDA* to Forest, Hills, and Lake Map

We will now outline the application of the basic IDA* algorithm to the pathfinding example introduced earlier.

Iteration 0:
threshold = 2 (heuristic estimate for start node)

pass 0:
   open [start2]; closed []
Expand Start2:

pass1:
   open [c4, b6, a4]; closed[start2]

The initial threshold has been exceeded for all nodes in the open list. We now start a new iteration with the threshold reset at the lowest f value in the open list—4.

Iteration 1: Threshold 4.
pass 0:
   open [start2]; closed []

Expand Start2:
pass1:
   open [c4, b6, a4]; closed[start2]

Note at this point that no re-ordering of the open list is performed, as is required in A*. If a node has an f value associated with it that is less than or equal to the current threshold then it can be chosen for expansion.

We're using DFS so child nodes are placed on the left of the open list; the leftmost node is then chosen for expansion.

Expand c4:
pass 2:
   open [e6, d8, b6, a4]; closed [start2, c4]

Now as node a4 is the only node on the list with f value less than or equal to the current threshold (4) we expand this.

pass 3:
   open [g5, f6, e6, d8, b6]; closed [start2, c4, a4]

As all nodes exceed the current threshold (4) this iteration is at an end. The threshold for the next iteration is again the lowest cost of the nodes on the frontier (the open list), in this case the value associated with node g (5).

Iteration 2: Threshold 5.

Pass 0, pass 1, and pass 2 as for iteration 2.
pass 3:
 open [g5, f6, e6, d8, b6,]; closed [start2, c4, a4]

Expand g5:
pass4:
 open [h7, j5, f6, e6, d8, b6]; closed [start2, c4, a4, g5]

Expand j5:
pass5:
 open [goal5, k7, l7, h7, f6, e6, d8, b6,]; closed [start2, c4, a4, g5, j5]

As goal5 is the next state in line to be expanded we have arrived at the goal, and with a guaranteed minimal cost.

Although we have applied IDA* successfully in this example to our pathfinding problem as an illustrative example of its application, we note it is not normally applied in its standard form to pathfinding problems on grids, as the depth-first-search element of IDA* causes problems where there are multiple paths to each node (Korf and Zhang 2000). However, IDA* has been applied with great success to sliding-piece puzzles. Indeed, in its original formulation in Korf (1985) it was claimed that:

> IDA* is the only known algorithm that [could] find optimal paths for randomly generated instances of the Fifteen Puzzle with practical time and space constraints...

7.6   Search in Two-Person Adversarial Games

Search in two-person adversarial games, such as chess, or draughts, or noughts and crosses poses a problem not encountered in puzzles such as the 8- or the 15-puzzle, namely the presence of an opponent behaving in a hostile and generally unpredictable manner. Essentially half of all of the moves that are made in the game are beyond the player's control. On the face of it initially this issue appears a daunting and even near insurmountable problem.

Von Neumann states the problem well in his 1928 paper "Zur Theorie der Gesellschaftspiele" ("On the theory of games of strategy") (von Neumann 1928)

> n players S_1, S_2, ..., Sn are playing a given game of strategy,...,How must one of the participants, S_m, play in order to achieve a most advantageous result?

In our discussion we will confine our consideration to two-player games in which the element of chance does not play a part, such as chess and draughts, but not, for example, poker or backgammon.

7.6.1   The Minimax Procedure

However, of course, in one sense chance *does* play a major part in games such as chess, as we are unsure after each move what our opponent will do next. To eliminate this complicating factor, we could make the simplifying assumption that our opponent is playing with exactly the same strategy as ourselves but *in the opposite direction*. So, a move that is worth +*v* to us is worth −*v* to our opponent. Historically the two opponents are called MAX and MIN. MAX (us) is trying to MAXimise her advantage in order to win the game, while MIN is trying to MINimise MAX's advantage. This is the basis of the minimax algorithm.

7.6.1.1   Exhaustive Minimax

The overall procedure is now quite straightforward, when applied to a simple game. We generate the entire search space for the game in question and label each leaf node: +1 (say) for a win for MAX, and 0 for a win for MIN. We now propagate these values up through the search tree, where at each level we choose the maximum value of a node's children if MAX is to play, and we choose the minimum value of its child nodes if MIN is to play. Now, finally, at the top of the tree with MAX (generally) to play, we choose the child node with the maximum value as our move.

A simple example should illustrate this procedure, known as *minimaxing*, for obvious reasons. We will use as our example a simple game known as *Grundy's game,* which is a variant of the game of *Nim*, itself estimated to be one of the oldest games in existence (Berlekamp et al. 2001). Nim is also generally reckoned to be the world's first computerised game, developed by Ferranti and demonstrated on May 5th, 1951 at the Festival of Britain as described in Donovan and Garriott (2010).

Two players start with a pile of objects (matches, stones, pens, etc.), and at each turn must divide one of the piles that are generated (or just the first pile in the case of the first move) into two further piles with a *different* number of objects in each pile. So, for example we can divide a group of six tokens into 1–5, or 2–4, but not 3–3. Using *normal play*, the first person unable to make a legal move loses.

We will look at the application of minimax to the five and six token versions of this game. In each case it is a simple matter to generate the full search tree. Then, starting at the leaf nodes we pass the correct values back up the tree, depending on whether we are at a MAX or a MIN layer. Values generated at leaf nodes are shown in circles, values generated by the minimax algorithm are shown in shaded circles (Fig. 7.6). It is clear in both cases that MAX is guaranteed a win given correct play. In the five-token case the left hand (1–4) option, and in the six-token case the right hand (4–2) option seals MINs fate. In the six-token case a win is still possible for MAX if he takes the left-hand (1–5) route, however given correct play by MIN the game will be lost.

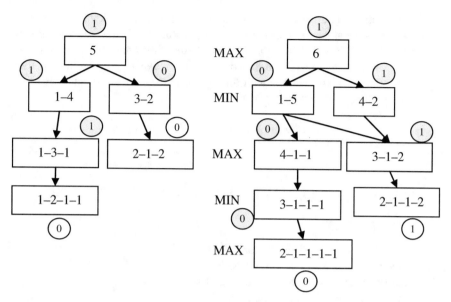

Fig. 7.6 Application of the minimax algorithm to the five (left) and six (right) token versions of Grundy's game

7.6.1.2   Depth-Limited Minimax

The observant reader may at this point have noticed a glaring issue arising in the application of minimax to games of any significant degree of complexity. It will, of course be impossible to generate more than a fraction of the search tree for complex games such as chess and Go. Some alternative methodology must be found. One possibility is to use the standard minimax algorithm but to a *limited depth*, generally an even number of *ply* (where we define a ply as a single move by either player). Of course, we then run the risk of not being able to see "over the horizon" but there are steps that can be taken to reduce problems in this area.

Then, because we are not dealing with leaf nodes that provide us with definite information as to a win or a loss, the following question arises: how do we assign values to these newly generated leaf nodes? We will need some *heuristic estimate* of the relative worth of these individual nodes. We will look at an example now, using the simple game of noughts and crosses, which should illustrate this process clearly.

7.6.1.3   Two-Ply Minimax Applied to Noughts and Crosses

Figure 7.9 illustrates the application of two-ply minimax to the game of noughts and crosses (tic-tac-toe). This game will be familiar to most readers; two players play alternately as X and O on a 3 × 3 grid; the first player to get "three in a row" in a horizontal, vertical, or diagonal direction wins.

In formulating a heuristic for this problem, it is clear that it is advantageous for a player if they have at least one of their "pieces" in a line unobstructed by all of their opponent's pieces. It is also clear that it is even better if they have two of their pieces in such a row. On the other hand, of course it is disadvantageous to them if their opponent holds similar positions with their pieces. Our heuristic function $E(n)$ for state n might then look something like this:

$$E(n) = (M(n) - O(n)) + 100(X_{win} - Y_{win})$$

where

$$O(n) = 3A + B, \text{ and } M(n) = 3C + D$$

and using

A = number of lines with two O's (and no X's)
B = number of lines with one O (and no X's)
C = number of lines with two X's (and no O's)
D = number of lines with one X (and no O's)
X_{win} = lines with three X's
Y_{win} = lines with three O's

e.g. for the position in Fig. 7.7
    $A = 0, B = 1, C = 1, D = 3$ and $E(n) = +5$

This heuristic simply rewards players based on the number of pieces they have in a line. Given the starting position shown in Fig. 7.9, X has four different options, if we ignore symmetrically equivalent positions. For example, in Fig. 7.8 the two illustrated positions are deemed to be equivalent:

For each of X's four choices we generate all possible moves for O (again ignoring symmetrically equivalent positions). For clarity in Fig. 7.9 we show O's responses to just two of X's possible moves. For the other two moves we just show the backed-up responses—the reader is invited to verify that these values are correct. For example, for X's leftmost move, three responses are possible from O, yielding values of -1, 1, and 0 respectively. We choose the lowest of these values (-1) to pass up to the

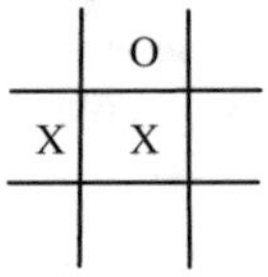

Fig. 7.7 Two X's in a row

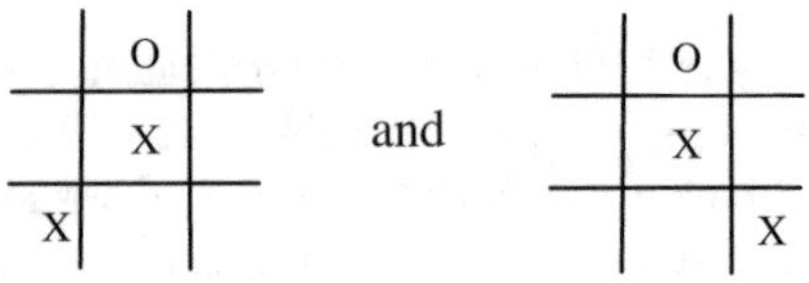

Fig. 7.8 Two symmetrically equivalent tic-tac-toe positions

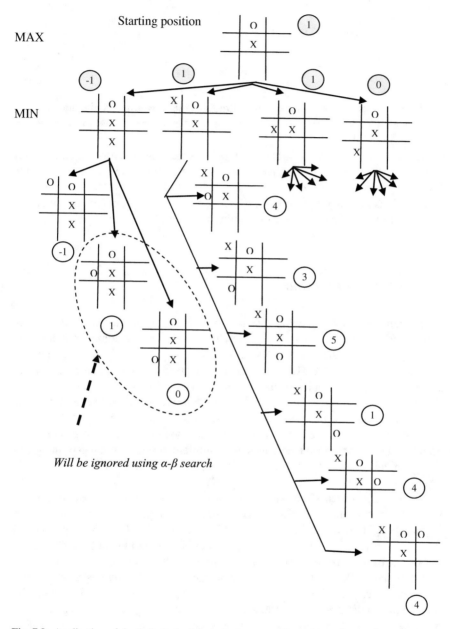

Fig. 7.9 Application of depth-limited minimax to the game of noughts and crosses

layer above as MIN is to play. Once the four backed-up responses are generated, we (as MAX) then choose the move with the highest value. In this example we see that either of the two centre moves, which have equal value, may be chosen.

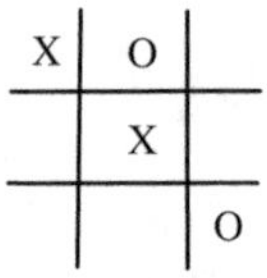

Fig. 7.10 MAX plays
leftmost

If we now assume MAX makes the leftmost of these two moves (the one in which we have elaborated O's six possible responses), after two moves we would expect to be as in Fig. 7.10.

This leaves X with five possible responses. It is clear that a subsequent move by X in either of the two leftmost squares will result in a win, as two possible winning lines are generated, only one of which can be blocked by O. One of these two combinations is certain to be chosen by X again using two-ply minimax (the reader may care to verify), thus irrevocably sealing O's fate.

7.6.2 The α-β Procedure

While minimax may appear an effective procedure on the face of it, there is of course, lurking in the background, as we have mentioned, the combinatorial explosion issue; that is, what happens when we expand the depth of our search beyond 2-ply to 6-, 8-, 10-, or even 12 ply. The numbers quickly become quite enormous for games of moderate complexity such as chess, even for modern computers. We can then use a technique called *α-β search*, or *α-β pruning*, which has been, in a sense, the de facto standard for adversarial search for some time, although fairly recent developments have shifted the emphasis towards various variants of Monte Carlo search and other advanced techniques we will briefly discuss later. α-β pruning was the technique used in the Deep Blue chess-playing machine that defeated Garry Kasparov, the world chess champion, in 1997.

While a little intricate in its computational detail, the basic principle of α-β search is easy to understand. If you play chess, or some similar game, you probably employ in your thought processes, either consciously or unconsciously, a technique such as this to generate your next move. A simple example should illustrate the procedure; full details of the algorithm are given in many other texts.

α-β pruning is so called because it generates the same move as a fixed-depth minimax search but with on average half the computational effort. It does this by pruning the minimax search tree and effectively eliminating those branches that will never be taken, and so do not need to be evaluated.

To do this α-β operates in a DFS fashion. So, let us illustrate the procedure with a simple example based on our noughts and crosses example from above.

7.6.2.1   α-β Applied to Noughts and Crosses

Let us assume that in the DFS that α-β search employs the first and second "columns" of the search are exchanged. Now, unlike standard depth-limited minimax, which generates all of the estimated values at a particular selected depth, α-β search just generates a subset of these values, depending on the particular search space in question, automatically eliminating those nodes incapable of producing a better result than that already evaluated.

So, getting back to our example, we do a depth-limited DFS, in this case initially generating a MIN value of 1; this is passed up to its parent node, and also to its grandparent. Moving on to the next "column" of our DFS we hit a -1 as our initial leaf node. Clearly this value, or any value lower, will be selected by the minimising layer above. However, as a result of this fact, there is no chance that the maximising layer above (our move) will then select this move over the one examined previously. So, there is no point in even evaluating the heuristics for the other two move options here, thus saving in computational time and effort.

This basic idea is at the heart of α-β pruning which was at the core of the algorithmic processes that sealed Garry Kasparov's fate and forever changed the way that the general public perceived the computer—not just as an anonymous number-crunching device with some business and entertainment applications, but as a potential rival to man's intellectual dominance on Earth; after all, if a computer can beat the world chess champion, what intellectual domain is still sacred?

7.7   Machine Learning

In general learning is a relatively permanent change in behaviour brought about by experience
—Jacek Zurada (1992). *Introduction to Artificial Neural Systems*

...the machine...which can learn and can make decisions on the basis of its learning, will in no way be obliged to make such decisions as we should have made, or will be acceptable to us. For the man who is not aware of this, to throw the problem of his responsibility on the machine, whether it can learn or not, is to cast his responsibility to the winds, and to find it coming back seated on the whirlwind.
—Norbert Wiener (1954). *The Human Use of Human Beings.* Houghton Mifflin Company, New York.

We can broadly define machine learning as the ability of a system to absorb information from its environment without requiring an external intelligent agent to "program" it (Judd 1990). Learning invariably involves *change*. How much change, and at what intervals based on what criteria are some of the important issues involved in this process of change. Other important issues to consider are the level of feedback (if any) given to the learning agent, and the mechanism for embedding domain knowledge within the search mechanism. In this section we will now briefly explore each of these different topics and give a number of illustrative examples of machine learning applications. ●

7.7.1   Learning Systems

Learning is generally targeted as a core component of AIS, systems that lay claim to AGI (*Artificial General Intelligence*) abilities, and for which we may also use the term *Advanced Artificially Intelligent Entities* (A^2IEs). Learning is also at the core of many data mining applications. Samuel put it very well in his seminal paper written over 50 years ago on training a computer to play the game of checkers (Samuel 1959).

> We have at our command computers with adequate data-handling ability and sufficient computational speed to make use of machine-learning techniques, but our knowledge of the basic principles of these techniques is still rudimentary. Lacking such knowledge, it is necessary to specify methods of problem solution in minute and exact detail, a time consuming and costly procedure. Programming computers to learn from experience should eventually eliminate the need for much of this detailed programming effort.

Another key aspect of useful learning systems is the ability to *generalise.* In the case of a *supervised* learning system (learning with a teacher) this implies that we are able to apply the general principles and patterns obtained in the training set to other, unseen instances. Clearly this is a very important feature for any system dealing with complex, real-world, noisy data. This generalisation ability is also known as *induction*, and brings with it a number of important issues that need to be addressed. Foremost among these is the issue of *inductive bias*. Because, for real-world complex and interesting problem domains we can only present the learning system with a very small fraction of the possible range of experiences in the problem domain, the selection criteria used to determine this set may introduces biases that may skew the ability of the learning agent to deal with previously unseen instances.

A major factor in the categorisation of machine learning algorithms is the level of feedback provided to the algorithm employed. Three broad types of learning can be identified in both natural and artificial systems, based on the level of feedback made available to the learning agent. These are:

- supervised learning,
- unsupervised learning, and
- reinforcement learning.

Supervised learning broadly corresponds to learning with a teacher present; unsupervised learning corresponds to learning without such a teacher available; reinforcement learning uses a process of trial-and-error learning.

We may further categorise learning systems into the broad representational systems used to encode knowledge in the problem domain. These are

- Symbol-based

- Neural network, and
- Evolutionary systems

Symbol-based learning systems represent data explicitly in symbolic form, and the learning algorithm attempts to discover patterns in this data and to form generalisations. The neural network approach (sometimes called the *connectionist* approach), based broadly on learning in animal and human brains, typically represents data as the interconnection strengths between large numbers of simple processing units. Learning occurs typically by the alteration of these connection strengths (and possibly also the network topology) in response to repeated applications of the training data. The evolutionary approach, based broadly on the evolutionary process in nature, typically encodes candidate solutions to a problem domain in string form, a number of these strings then form a population. The performance of each individual in the population is then tested using a fitness function and individuals with a higher fitness are allowed transit to the next generation either unaltered, or slightly modified, or combined in some fashion with another individual from the current population.

These categorisations (feedback level and representational system) can then be combined in a variety of different fashions. For example, a number of deep learning applications use a reinforcement learning approach which is embedded in a neural substrate. Evolutionary learning can, in a certain sense be viewed as a category of reinforcement learning, if the fitness function used only evaluates an individual's performance after a long sequence of actions as is typically the case in an evolutionary robotics application. Neural networks can employ supervised, unsupervised, or reinforcement learning paradigms.

To illustrate these different machine learning paradigms, we will initially look at symbol-based learning algorithm called *ID3*, which has formed the basis for many other powerful symbol-based learning applications. We will then explore the area of neural network learning, in particular a supervised learning paradigm called the perceptron learning algorithm, which, in a much-simplified form, forms the basis for many of the powerful deep learning classifiers in use today. We will then explore the area of reinforcement learning, looking at a particular algorithm called temporal difference learning, the idea of which is at the core of many powerful reinforcement learning algorithms, such as *Q-learning*. Finally, we will look to evolutionary systems, and in particular the genetic algorithm, which is used in a wide variety of evolutionary applications.

We will look at four different examples to illustrate these diverse machine learning approaches. One example uses a supervised neural network to learn a Boolean function, such as those discussed in Chap. 2. The other three examples are game playing applications, the use of symbol-based learning to aid an NPC in a computer game in their mushroom foraging expeditions, using reinforcement learning to create an effective tic-tac-toe player that can adapt to its opponent's playing strategy, even if this changes over time, and the use of an evolutionary robotics approach to generate ball-kicking behaviour in simulation, which can be transferred successfully to a real humanoid robot.

7.7.1.1 Learning, Optimisation, and Search

You can search without optimising. Optimisation need not involve search (for example mathematical optimisation). Learning may involve both search and optimisation. You can search without learning (don't store long-term the results of your search). You can learn without searching (just add new exemplars to a database—however *retrieval* may involve search).

We might note, for example, that in a genetic algorithm (GA) evolving robot motions (say, ball-kicking behaviours), a *search* through the player search space generally leads to an *improvement* in the individual robot soccer players in each succeeding generation. This, in fact, is the whole goal of the search in the first place. So, the question arises—where does search stop and learning start?

A learning process in general implies an individual improving its performance over time. In GAs, however we have a multiplicity of individuals evolving over generations. For a generational GA perhaps you could argue that each generation represents the behaviour of a particular agent (using the information from the fittest genome to drive the robot, in an ER context) and, of course, in an elitist GA we (hopefully) then have a continuously improving "individual"—can we not then characterise this as a learning situation?

An argument might be made that as this concept of an "organism" changes fairly radically form one generation to the next, there is not such a sense of continuity as one might have in, say, reinforcement learning or backpropagation-based neural network learning. This brings into focus the issue of how important is the notion of gradual change to the concept of learning—e.g. in decision tree learning with ID3 we need to run the whole algorithm again in order to accommodate another training instance.

7.7.1.2 Lifelong Learning

A current area of major research effort is in the area of "lifelong learning"—that is learning as humans do over the entire lifetime of the agent, with new learned behaviours, rather than overwriting previously learnt ones, instead augmenting and improving these behaviours, or, indeed, adding completely new abilities to the learning agent. The whole area of learning in animals—especially humans and the higher primates—is predicated on the fact that early experience greatly outweighs the later. In fact this area of lifelong learning is now considered so important that the US DARPA (Defence Advanced Research Projects Agency), the same body that funded the development of the Internet and much of the early research into self-driving cars; has recently instigated a major research effort in this general area.

7.7.2 The ID3 Decision Tree Induction Algorithm

The ID3 (*Iterative Dichotomiser 3*) algorithm as proposed by Quinlan in 1986 allows us to induce general concepts from discrete examples (Quinlan 1986). ID3 represents concepts as *decision trees*. A decision tree is a representation allowing for the classification of an individual by testing values for particular properties. In classifying an individual it may not be necessary to use all the properties available The size of the tree required to classify a particular set of examples will generally vary according to the order in which they are tested. ID3 assumes that the simplest decision tree that covers all of the training examples will be the most efficient for the future classification of unseen instances of the population.

This is based on the principle "Occams Razor" proposed by William of Occam in 1324

> It is vain to do with more what can be done with less... Entities should not be multiplied beyond necessity

Or in more modern language — "keep things as simple as possible".

7.7.2.1 Information Theoretic Test Selection

Each separate property can be seen as contributing a certain amount of information to its classification. ID3 operates by calculating the information gain by placing each property at the root of the current subtree and then choosing the property with the highest information gain.

Claude Shannon in 1948 (Shannon 1948) provided a means of numerically calculating the information content of an individual message I(M) (in bits) as

$$I(M) = \sum_{i=1}^{n} -p(m_i) \log_2(p(m_i))$$

For example, the information content of the flip of a fair (unbiased) coin is

$$= -\frac{1}{2} \log_2\left(\frac{1}{2}\right) - \frac{1}{2} \log_2\left(\frac{1}{2}\right)$$
$$= 1 \text{ bit}$$

[Note: for any bases b and k,

$$\log_b(x) = \frac{\log_k(x)}{\log_k(b)}$$

so, we can convert easily from base 10 (or any other base) to base 2. So, for example, to convert from base 10 to base 2 we simply divide by 0.301.]

However, if the coin has been biased to come up heads (or tails) 60% of the time we have

$$I(biasedtoss) = -\frac{6}{10} \log_2\left(\frac{6}{10}\right) - \frac{4}{10} \log_2\left(\frac{4}{10}\right) = 0.97 \quad \text{bits}$$

so, the information content is less, as we would expect.

7.7.2.2 ID3 Example

Table 7.3 summarises the experience to date of an NPC (non-player-character) based on its mushroom gathering forays, in a role-playing game. We now want to use ID3 to generate the simplest decision tree that fit the data presented, in the expectation that this tree will be the best at classifying new mushroom encounters. This example is adapted from that given in Luger (2009).

Our first task is to get an estimate of the overall information content of the table above. If we assume that all the examples in the table occur with equal probability, we have

$$p(poisonous) = 6/14$$
$$p(edible) = 3/14$$
$$p(nutrituous) = 5/14$$

Hence the information content of the distribution in this table (and hence any associated decision tree covering these examples) is

Table 7.3 Data collected from a series of mushroom gathering forays

No.	Type	Taste	Colour	Size
1	Poisonous	Bitter	Red	Small
2	Poisonous	Bland	Red	Medium
3	Edible	Bland	Brown	Medium
4	Poisonous	Bland	Brown	Small
5	Nutritious	Bland	Brown	Large
6	Nutritious	Bland	Brown	Large
7	Poisonous	Bitter	Brown	Small
8	Edible	Bitter	Brown	Large
9	Nutritious	Pleasant	Brown	Large
10	Nutritious	Pleasant	Red	Large
11	Poisonous	Pleasant	Red	Small
12	Edible	Pleasant	Red	Medium
13	Nutritious	Pleasant	Red	Large
14	Poisonous	Bitter	Red	Medium

$$I(Table) = -\frac{6}{14} \log_2\left(\frac{6}{14}\right) - \frac{3}{14} \log_2\left(\frac{3}{14}\right) - \frac{5}{14} \log_2\left(\frac{5}{14}\right) = 1.531 \text{ bits}$$

Now we know the gain in information by making a test at the root of the current tree is equal to the total information in the tree less the information required to complete the classification after making this test. The amount of information required to complete the tree can now be calculated by multiplying the information content of each subtree by the % of examples present in that subtree and summing over these products.

Assume we have a set of training instances, C. If we make property P, with n values, the root of the current tree, this will partition C into subsets. The information expected to be needed to now complete the tree is:

$$E[P] = \sum_{i=1}^{n} \frac{|C_i|}{|C|} I[C_i]$$

and

$$gain(P) = I[C] - E[P]$$

If we make size the property tested for the root of the tree, we have:

$$C_1 = \{1, 4, 7, 11\} - small$$
$$C_2 = \{2, 3, 12, 14\} - medium$$
$$C_3 = \{5, 6, 8, 9, 10, 13\} - large$$

and the expected information needed to complete the tree is:

$$E[P] = \sum_{i=1}^{n} \frac{|C_i|}{|C|} I[C_i]$$

where

$$|C| = 14, |C_1| = 4, |C_2| = 4, |C_3| = 6$$

so

$$E[size] = \frac{4}{14} I[C_1] + \frac{4}{14} I[C_2] + \frac{6}{14} I[C_3]$$

Now we know:

$$I(M) = \sum_{i=1}^{n} -p(m_i) \log_2(p(m_i))$$

so:

Poisonous   Edible   Nutrituous

$$I(C_1) = -\frac{4}{4} \log_2\left(\frac{4}{4}\right) - \frac{0}{4} \log_2\left(\frac{0}{4}\right) - \frac{0}{4} \log_2\left(\frac{0}{4}\right) = 0$$

$$I(C_2) = -\frac{2}{4} \log_2\left(\frac{2}{4}\right) - \frac{2}{4} \log_2\left(\frac{2}{4}\right) - \frac{0}{4} \log_2\left(\frac{0}{4}\right)$$

$$I(C_2) = -\frac{2}{4}(-1) - \frac{2}{4}(-1) = 1.0$$

$$I(C_3) = -\frac{0}{6} \log_2\left(\frac{0}{6}\right) - \frac{1}{6} \log_2\left(\frac{1}{6}\right) - \frac{5}{6} \log_2\left(\frac{5}{6}\right)$$
$$= 0.431 + 0.219 = 0.65$$

so

$$E(size) = \frac{4}{14} \times 0.0 + \frac{4}{14} \times 1.0 + \frac{6}{14} \times 0.65$$
$$= 0.564 \text{ bits}$$

that is, the estimated amount of information required to complete the tree after placing size as the root node is 0.564 bits. Now we know:

$$gain(P) = I[C] - E[P]$$

that is: the estimated gain in information expected after placing size as the root node is the total information in the tree (or subtree as we move down the decision tree) minus 0.564 bits.

As we have shown the estimated information content of the distribution in this table is 1.531 bitsso, gain(size) = 1.531 − 0.564 = 0.967 bits.that is: the expected gain in information by placing size as the root of the tree is 0.967 bits.

We now need to do the same thing with the other two properties: colour, and taste. Finally, we will pick the property with the highest information gain for the root of the tree, in line with the principle of ID3; that is to come up with the simplest decision tree that will fit the training set presented. Now if we make colour the property tested for the root of the tree, we have:

$$C_1 = \{1, 2, 10, 11, 12, 13, 14\} - red$$
$$C_2 = \{3, 4, 5, 6, 7, 8, 9\} - brown$$

and the expected information needed to complete the tree is:

$$E[colour] = \frac{7}{14}I[C_1] + \frac{7}{14}I[C_2]$$

Poisonous   Edible   Nutrituous

$$I(C_1) = -\frac{4}{7}\log_2\left(\frac{4}{7}\right) - \frac{1}{7}\log_2\left(\frac{1}{7}\right) - \frac{2}{7}\log_2\left(\frac{2}{7}\right)$$
$$= 0.461 + 0.401 + 0.516 = 1.378$$

$$I(C_2) = -\frac{2}{7}\log_2\left(\frac{2}{7}\right) - \frac{2}{7}\log_2\left(\frac{2}{7}\right) - \frac{3}{7}\log_2\left(\frac{3}{7}\right)$$
$$= 2 * 0.516 + 0.524 = 1.556$$

so

$$E(colour) = \frac{7}{14} \times 1.378 + \frac{7}{14} \times 1.556 = 1.467 \text{ bits}$$

and gain(colour) = 1.531 − 1.467 = 0.064 bits
     Similarly, we can show that

$$gain(taste) = 0.266 \text{ bits}$$

As size has been shown to give the highest information gain, ID3 now selects it as the root of the tree. This analysis is now applied recursively to each subtree until the overall tree is completed.

For the small size case, all of the training set examples corresponding to this case are poisonous, so we label a leaf node with this class (poisonous).

For the medium size case, using an analysis similar to that above, we obtain:

$$gain[taste] = I(med) - E(taste) = 1 - \frac{1}{2} = 0.5$$

and

$$gain[colour] = I(med) - E(colour) = 1 - 0.689 = 0.311$$

So, taste is the next property to be checked, given medium size, as it presents the highest information gain (0.5 bits). Bad taste now translates to poisonous, good taste translates to edible (one instance of each). There are two examples of unknown taste, colour red/brown distinguishes between poisonous/nutritious.

Finally taking the large size category we have objects numbered

$$\{5, 6, 8, 9, 10, 13\} - large - size$$

Taking taste as the next property to test, we have

$$C_1 = \{8\} - bitter$$
$$C_2 = \{5, 6\} - bland$$
$$C_3 = \{9, 10, 13\} - pleasant$$

$$E[taste] = \frac{1}{5}I[C_1] + \frac{2}{5}I[C_2] + \frac{3}{5}I[C_3]$$

giving

$$E[taste] = \frac{1}{5} \times 0 + \frac{2}{5} \times 0 + \frac{3}{5} \times 0 = 0$$

This result says that it is not expected that any information will be required to complete this subtree given that taste is chosen as the next property to test. Clearly this must be the maximum gain, so given large size, the next property to test is taste. This test will complete the decision tree, correctly classifying all of the examples, as shown in Fig. 7.11. It is also the simplest decision tree that will perform this task, which we also hope will be the most accurate at classifying hitherto mushroom instances.

So, for example, if in the future our NPC comes across a mushroom which is small, brown, and bitter—an example which it has not hitherto encountered, it can conclude from the ID3 generated decision tree that the mushroom is, in all likelihood, poisonous. However this is not guaranteed, since at its core ID3 is a heuristic algorithm because as pointed out in Millington (2009) so while the simplest decision tree as generated by ID3 is in a high percentage of cases demonstrated empirically to be the best for classifying unseen instances, this does not always prove to be the case.

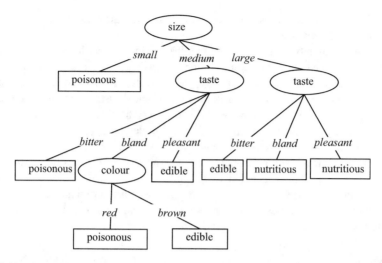

Fig. 7.11 The decision tree generated by the ID3 algorithm for the mushroom-gathering example

7.7.2.3   ID3 Variants

The ID3 algorithm has inspired many variants and new versions, including *ID4* (Fisher 1987) and *C4.5* (Quinlan 1993).

ID4 added incremental learning capabilities to the basic ID3 algorithm, this allows for adding the information gained from new training instances to an existing ID3 generated decision tree. C4.5 further enhanced the capabilities of ID3 by adding the ability to deal with incomplete data points and to handle continuous as well as discrete features. It also seeks to address the problem of over-fitting using a novel pruning algorithm. C4.5 is the most popular data mining algorithm according to a poll taken in 2006 by the International Data Mining Conference (Witten et al. 2016).

7.8   Introduction to Neural Computation

The Navy revealed the embryo of an electronic computer today that it expects will be able to walk, talk, see, write, reproduce itself and be conscious of its existence. Later perceptrons will be able to recognise people and call out their names and instantly translate speech in one language to speech and writing in another language, it was predicted.
—The New York Times (July 8 1958) "New Navy Device Learns by Doing"

Neural networks come in two basic types—natural and artificial. The natural kind we have had many millions of years' experience in dealing with; with the artificial types only a few decades. The natural type may be further subdivided if, like Descartes (1837), we consider there to be a fundamental difference between man and the other living species that inhabit this planet. This view, incidentally, is nowadays far from being universally accepted. We may then visualise three basic categories, human neural networks, non-human biological neural networks, and artificial neural networks.

Human neural networks (brains) form a fascinating subject for study, one reason being that they are the only machine (if one accepts they are a machine of sorts) that we can look at from the inside out, so to speak, as well as from the outside in. That is, as well as studying the gross and more minute anatomical structure of the brain (neurons, synapses, etc.), which we will shortly briefly discuss, we may also gain some knowledge from our own perception and understanding of the world around us, our consciousness for want of a better word. The human brain is also of intense interest because it is, without question, one of the most complex objects known to man. While studies have enabled us to understand some of the basic structural aspects of the brain (Eccles 1973), by far the greater portion remains to be explored. The nature of consciousness is a fascinating subject and the subject of much recent intense research; we address this topic further in Chap. 6.

The study of the second form of neural network—non-human biological neural networks, is of interest because many things may be learnt from their examination that would not be possible on human beings for ethical reasons. The third form of neural network, the artificial neural network, is the type we will discuss now.

Instead of attempting to simulate the coarser aspects of living entities, i.e. those with biological neural networks, our simulation will be at the level of the neuron and also at the level of the connection together of a relatively small number of neuron-like elements. One reason for our interest is that while modern-day digital computers are extremely good at calculations such as adding a few hundred 20-digit numbers together, (which human brains find very difficult), some tasks that the human brain carries out quickly and apparently without effort, such as the recognition of patterns, the solution of crossword puzzles, or difficult control problems in the face of uncertainty, can prove extremely time-consuming and indeed, until fairly recent times, sometimes even impossible on the digital computer.

This is in spite of the fact that while modern computers' instructions can be executed in nanoseconds, the basic neuronal cycle (the time for the neuron to operate) is on the order of milliseconds. It is accepted that one of the main driving factors behind the processing ability of the brain is the massive degree of parallelism employed there. This is supported by the fact that some 10^{10} neurons are contained in the brain, each of which may typically be connected to 10^4 others. So, at any instant in time an enormous number of these neurons may be operating in parallel and altering their individual states in response to each other.

What of the processing carried out in each individual neuron (sometimes called a processing element or node if speaking of an artificial neuron)? The actual behaviour of the individual neuron is known to be complex. However, as we are interested in the behaviour of the neural network as a whole we can simplify our model of the basic neuron in the hope that not too much functionality will be lost.

Input signals to the neuron come either from the environment or from the outputs of other processing elements, and may be taken to form an input vector $A = (a_1, a_2, a_3, \ldots, a_n)$ for a processing element with n inputs, where a_i is the input associated with the ith processing element. Associated with each connected pair of processing

Fig. 7.12 A simple model neuron

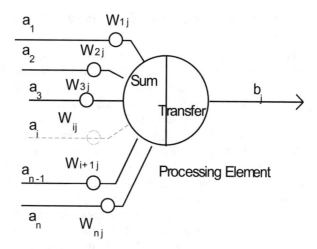

elements is an adjustable value called a weight. The collection of weights into the jth processing element then forms a vector $W_j=(W_{1j}, W_{2j}, W_{3j},..., W_{nj})$.

In addition, there is sometimes an additional parameter θ_j modified by the weight W_{0j}. This can be regarded as an internal threshold value that has to be exceeded for the activation of the processing element. This corresponds broadly to the neural threshold present in some natural neurons. A typical computation performed by processing element j is as follows:

$$b_j = f\left(\sum_{i=1}^{n} a_i W_{ij} - W_{0j}\theta_j \right)$$

Here b_j is the output of processing element j and f is the activation function which this value is passed through. Activation functions include linear functions, signum functions, threshold logic elements, and sigmoidal non-linearities. More complex processing elements may include temporal integration or some other time dependant behaviour and may use some other operation than summation at the input stage. See Fig. 7.12 for a pictorial representation of a simple model neuron.

The advantages of the neural net approach go beyond the high processing speeds allowed by massive parallelism. In most neural network models there is no single centralised memory. Instead the memory of the system is distributed throughout the system and generally assumed to be contained in the connection weights. This localised and distributed memory and processing architecture gives a more robust and fault-tolerant architecture than von Neumann sequential computers, where the malfunction of a single gate or connection may typically cause a system crash.

7.8.1 History of the Development of Artificial Neural Systems

The first researchers to formulate the fundamentals of neural computing were McCulloch and Pitts in the early 1940s (McCulloch and Pitts 1943). The neuronal model they used was that of a binary device having a fixed threshold. The neuron can receive input from both excitatory and inhibitory synapses. All of the excitatory synapses have the same weights. The basic operation of the neuron is simple; if no inhibitory synapses are active the excitatory inputs are summed and compared with the threshold value. If the sum exceeds the threshold the neuron becomes active for this time step (the time step being based on the physiologically observed synaptic delay); if not the neuron remains inactive. If any inhibitory synapse is active the neuron will not become active. This simple model can be shown to implement a number of logical functions and can be regarded as laying the foundations for further developments; however the model was not widely implemented in hardware.

Moving on to 1949 Donald Hebb (1949) proposed a method for allowing neurons to learn, that is to change their connection weights in response to external stimuli. Then in the 1950s models of neural networks were set up on digital computers (Farley and Clark 1954; Rochester et al. 1956) and the first hardware neurocomputers were built and tested (Minsky 1954). Later on, in this decade a new neuron-like element called the *perceptron* was proposed by Frank Rosenblatt in 1958 (Rosenblatt 1958). This device, unlike some earlier models, had trainable connection weights and forms the basis for some of the neural systems in use today.

The perceptron was a pattern classifier capable of making generalisations of a limited nature and of categorising patterns in the presence of noise. A grid of photocells received the optical input; these photocells were randomly connected to association units. If the input from the photocells exceeded a certain threshold then the association units instructed response units to produce an output. While the perceptron worked well on certain patterns, because of the linear nature of the perceptron it could never recognise others. This defect was pointed out clearly in 1969, in Minsky and Papert's book *Perceptrons* (Minsky and Papert 1969) which has been recently re-published in 2017. After publication of this book work in this area came to a virtual halt, some would say as a direct result of this publication; others including Minsky and Papert (1988) would argue that because of the lack of adequate basic theories to explain the learning mechanisms involved, and why certain patterns could be recognised and others could not, the field was already on its last legs.

Research in the field died down then until the early 1980s. Notable exceptions include the *Adaline* or Adaptive Linear Combiner, the new learning procedure the Widrow-Hoff learning rule developed by Bernard Widrow and Marcian Hoff (Widrow and Hoff 1960), the *Neocognitron* (Fukushima 1980) for pattern recognition, which provided the inspiration for modern highly-successful convolutional neural networks, developments in the mathematical theory of neural networks (Amari 1977), and Grossberg and Carpenter's work (Grossberg 1980). The pace of research quickened from 1982 to 1986 with the publication of John Hopfield's

(1982) paper on a recurrent neural network architecture for associative memories, Kohonen's unsupervised learning networks for feature mapping into arrays of neurons (Kohonen 1982), and Barto et al.'s (1983) paper on the balancing of an inverted pendulum using simple neuron-like elements.

Then, late in 1986, came the publication of two volumes entitled *Parallel Distributed Processing* by James McClelland and David Rumelhart (Rumelhart and McClelland 1986). These texts showed how the limitations imposed by single-layer networks could be largely overcome by the use of multi-layer networks and how these networks could be effectively trained, even though the mathematical framework for the new training scheme had already existed for some time (Werbos 1974). From 1986 to the present the field of artificial neural computing has seen enormous growth with applications in many and varied areas from small tasks to large practical applications. The number of conferences and journals devoted to the field has grown at a rapid rate, very large-scale integrated neural network chips have been fabricated, and university courses at undergraduate and postgraduate levels have been set up in the field. However, even given this massive upsurge in interest much remains to be discovered and achieved, and most researchers would still contend today that in many areas our biological benchmark will not be bypassed for many years into the future.

7.8.2 Learning in Neural Networks

Learning is thought to occur in natural or biological neural networks as a result of modifications made to the way one cell is connected to another, at the synaptic junction. To get a clearer picture of the process Fig. 7.13 shows an enlarged and much-simplified picture of a neuron in the brain.

We see that attached to the main cell body, on the left, are several irregular filaments called dendrites. These filaments are usually less than a micron in diameter and branch out in a complex fashion. These dendrites accept the input to the cell via the synapses connected to them. These inputs may come via the axonal output of another neuron. The axon operates on electrical principles, unlike the dendrites, and forms the output channel for the neuron, this output taking the form of a sequence of

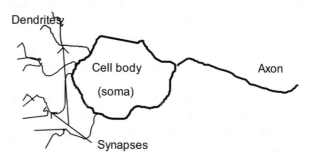

Fig. 7.13 Simplified biological neuron

rapid voltage spikes produced when the resting potential of the soma rises above a certain value. The connection from the axon of one cell to the dendrite of another is effected by the release of chemicals called neurotransmitters. Sufficient quantities of these neurotransmitters may alter the potential of the soma above the critical value, causing it to fire. It is also possible for an axon not to propagate a spike, but instead to produce a graded potential. However, this type of communication may only occur over short distances because of attenuation. This type of connection is thought to generally occur in the cerebral cortex which is assumed to be the seat of many high-level functions in the brain. Also, to complicate things, some synapses operate on electrical, not chemical principles.

Learning, then, in biological networks is thought to occur when changes are made to the connection between one cell and another, at the synaptic function. Two types of synapse are thought to exist: Type I and Type II. Type I synapses appear to be excitatory and in great abundance, Type II appear to be inhibitory and are fewer in number. By changing the connection at the synapse to allow the release of more or less neurotransmitters to the dendrite the effect of a signal on the affected axon can be altered, and the operation of the network altered slightly. Artificial neural networks also learn by altering the connections between neurons, a process known as changing the weight on a connection.

7.8.2.1   Reinforcement Learning

An interesting and early example of a machine that learnt by altering strengths between nodes is the MENACE (*Matchbox Educable Noughts and Crosses Engine*) for machine playing of the game of noughts and crosses. This system was developed by Donald Michie (Michie and Chambers 1968) in the days of expensive computer hardware, and consisted of a block of 288 matchboxes, each containing a number of beads, each matchbox corresponding to one of the possible board positions with which the player starting the game may be faced.

Learning occurs in each of the separate boxes which can be filled with coloured beads each colour corresponding to moves for different squares of the board. The decision of which move to make next is made by random choice of a bead from the box. Learning in this system is effected by, on the completion of a game, a system of reinforcements, i.e. increasing the probability of making those moves which led to a win, and decreasing the probability of making those moves which led to a loss.

This type of learning is known as reinforcement learning and we will look further at its application to the game of noughts and crosses shortly, using a slightly updated approach. One of the problems associated with reinforcement learning, and indeed learning in general; is the credit assignment problem (Minsky 1961). That is, given a favourable (or not so favourable) outcome, how to decide which decisions or actions deserve credit for improvements in the general performance of the system over time. The problem is especially difficult when you are dealing with a system with complex non-linear dynamics, and when feedback on performance occurs only at the end of a sequence of actions.

7.8.2.2   Supervised Learning

Another form of learning in very common use is the supervised learning paradigm. In this paradigm we assume that whenever an input is applied to the learning system, the desired response of the system is also provided (Fig. 7.14).

The distance, measured in some fashion between the desired and actual response of the neural network, is then employed as an error signal and is used to alter the network's weights. Typically, a set of input values and desired responses, called a training set, are repeatedly presented to the network and the network gradually learns the correct responses. This type of learning is very common and is used in many instances of neural network learning.

7.8.2.3   Unsupervised Learning

Another form of learning involves learning without direct supervision, also known as unsupervised learning. In unsupervised learning the desired response is not known, so no error feedback information can be provided to the network (Fig. 7.15).

A typical task performed by an unsupervised network is the clustering of inputs and finding boundaries between classes of input patterns. The network attempts to find regularities and patterns in the statistical structure of the set of inputs. Learning may not always be possible, however, as the different classes will not always be easily found.

It is sometimes said that unsupervised learning corresponds to learning without a teacher. As pointed out in Zurada (1992) this may not be the case, as some

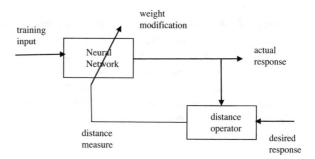

Fig. 7.14 Supervised learning

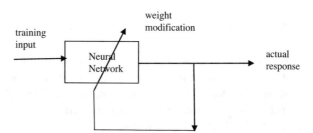

Fig. 7.15 Unsupervised learning

researchers would argue that no learning is possible without a teacher in some form, even if only to set the learner's goals. We can draw the classroom analogy of a foreign language class in a school. In this case, learning with supervision involves the teacher listening to each pupil speaking in the foreign language and correcting his speech after each sentence, and providing a general commentary and hints. Reinforcement learning might correspond to the situation where a student is given a 10 min passage to read out in the language; at the end of the passage the teacher just comments good, fair or bad, depending on the students' performance, not providing specific information on where, for example, mistakes might have been made. Unsupervised learning corresponds to the case where the student learns from a lecture, which has been recorded, but without any other feedback. The teacher gives the overall goal, i.e. to learn the foreign language, and some other general information but apart from that is not available for evaluative feedback.

7.8.2.4   Batch Learning

We may distinguish another mode of learning in neural networks, called "batch learning". This occurs when all the weights in the network are changed in a single training cycle, all the training information is provided together, and no feedback information is provided by the network itself in setting the weights. This is the system used in the Hopfield network (Hopfield 1982).

7.8.3   Scaling Up, and the Plasticity-Stability Dilemma

The topic of learning is of major importance in neural network research. Brains are extremely good at it; in particular they have overcome two of the major problems currently facing researchers interested in learning. The first of these problems involves the *plasticity-stability* dilemma, or how to construct a network that will be sufficiently stable to store learned data correctly, and yet flexible enough to allow the addition of new information at will. The second problem is that of how to scale up results obtained with small networks to larger networks without unacceptably large learning times.

7.8.4   The McCulloch-Pitts Neuron

The McCulloch-Pitts neuron was first proposed in 1943 (McCulloch and Pitts 1943) and was the first formal definition of an artificial neuron based on simplified biological models. It is of interest both from a historical perspective and from the fact that using even this highly simplified model it is possible to construct digital computer hardware for any required function.

Fig. 7.16 NAND gate
implemented using
McCulloch-Pitts neurons

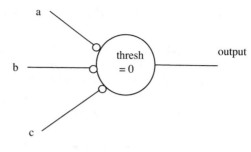

Fig. 7.17 Memory cell
implemented using
McCulloch-Pitts neurons

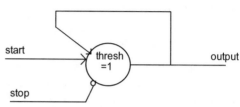

Time is divided into discrete intervals; in any interval the neuron can either be firing or in a quiescent state, corresponding to outputs 1 and 0. A neuron will fire at time $t+1$ if and only if at time t no inhibitory inputs are active and the sum of the excitatory inputs exceeds the threshold of the neuron. An alternative model also used by McCulloch uses a system where the cell fires if the difference between the excitatory and inhibitory inputs exceeds the threshold.

We will use the former model here. Using this simple basic structure, it is possible to construct two very useful networks, corresponding to a NAND gate and a memory cell in conventional digital systems. We may denote an excitatory input by an arrow, and an inhibitory input by a small circle (Minsky 1967). Figure 7.16 shows the implementation of a NAND function with three inputs. It is well known that any multivariable combinational function can be constructed using a combination of NAND (or NOR) gates.

We can also fabricate sequential logic activity by incorporating the memory cell shown in Fig. 7.17. Once this cell has an input on the *start* line it will continue to fire until it receives an input on the *stop* line; it remembers the last input it has received.

While it is interesting to note that we can duplicate the functionality of existing computer hardware using an artificial neural network constructed using these cells, this is not our primary purpose. Rather we wish to explore how massively interconnected nets of artificial neurons, not much more complex than those described here, operating in parallel, can learn complex tasks from experience.

7.8.5   Single- and Multi-layer Perceptrons

The McCulloch-Pitts neuron, described in the previous section, is interesting in that we have seen that it should be possible to construct quite powerful devices using combinations of these artificial neurons. However, no indication has been given as to how to create a device from these simple devices that will actually learn from experience. To do this we introduce a device with variable connection weights, which evolve using supervised learning: the *perceptron* (Rosenblatt 1958). Rosenblatt consciously modelled the perceptron on what he saw as some of the fundamental properties of intelligent systems, without in his own words

> becoming too deeply enmeshed in the special and frequently unknown conditions which hold for particular biological organisms.

The perceptron, as originally envisaged by Rosenblatt, was essentially a device for pattern recognition. It can be termed a single layer network if one does not count the input nodes as constituting a separate layer. In a multi-layer network neurons are arranged in successive layers starting from an input layer moving through an arbitrary number of hidden layers and ending at an output layer. Connections in a multi-layer network are generally assumed to exist only between adjacent layers. Layered networks may be contrasted with fully connected networks (e.g. the Hopfield net), in which each neuron can potentially affect, and be affected by, any other neuron.

7.8.5.1   Learning in Perceptrons

The overall learning paradigm of the single-layer perceptron involves initially setting random weights and thresholds. An input pattern is then presented to the network and the output is calculated by calculating the weighted sum of the inputs and passing this through a thresholding function. This output is then compared with the expected output (supervised learning) and the weights are altered accordingly. The learning rule used is a variant of Hebbian learning, so named because it was proposed in 1949 by Donald Hebb (1949) from his studies on natural (biological) networks and involves reinforcing active neuronal connections. The input to the network is binary-valued and may be treated as a vector $X_0, X_1, ..., X_n$. The output is then calculated from the input by the following formula:

$$y(t) = f\left[\sum_{i=0}^{n} W_i(t) \ X_i(t)\right]$$

where $f(x) = 1$ for $x > \ = 0$

$$f(x) = 0 \text{ for } x < 0$$

and where X_0 is always 1 with a weight equal to minus the threshold value θ. This value is known as the neuron's bias, or offset, and it avoids us having to include a threshold explicitly in the equation. Learning then occurs as follows. If the output of the neuron is correct, then

$$W_i(t+1) = W_i(t)$$

If the output value is 0, but it should be 1, then

$$W_i(t+1) = W_i(t) + X_i(t)$$

If the output value is 1, but should be 0, then

$$W_i(t+1) = W_i(t) - X_i(t)$$

This then is the basic algorithm; however, a number of modifications to it do exist. One commonly used modification allows the weights to be changed at a slower rate than is specified in the above equations by introducing a factor η into the weight adaptation equation; for example, if the output is 0 and should be 1 the revised equation is

$$W_i(t+1) = W_i(t) + \eta X_i(t)$$

7.8.5.2   Learning the NOR Function

We will now apply the basic version of the perceptron algorithm, just described, to learn the NOR function. As we saw in Chap. 2, the two-input NOR function produces an output which is TRUE (1) only in the case where both inputs are FALSE (0). The NOR function has the attraction that it is functionally complete, in that any other binary logic function can be generated using a combination of NOR gates (as with the NAND function).

As stated above, we will add an additional input, X_0 fixed at 1, the weight associated with this input is the bias for the neuron, allowing the threshold θ of the unit to be fixed at 0. The details of this learning process are given in Table 7.4.

We see that the weights have been altered several times by the repeated application of the training input to the initial values. The final values $W_0 = 0.2$, $W_1 = -0.7$, $W_2 = -1.1$ correctly compute the NOR function, as clearly demonstrated in Table 7.5.

Table 7.4 A perceptron learning the NOR function

X_0	X_1	X_2	W_0	W_1	W_2	Activation	Output	Expected output	Change weights?
1	0	0	0.2	0.3	−0.1	0.2	1	1	No
1	0	1	0.2	0.3	−0.1	0.1	1	0	Yes
1	1	0	−0.8	0.3	−1.1	−0.5	0	0	No
1	1	1	−0.8	0.3	−1.1	−1.6	0	0	No
1	0	0	−0.8	0.3	−1.1	−0.8	0	1	Yes
1	0	1	0.2	0.3	−1.1	-0.9	0	0	No
1	1	0	0.2	0.3	−1.1	0.5	1	0	Yes
1	1	1	−0.8	−0.7	−1.1	−2.6	0	0	No
1	0	0	−0.8	−0.7	−1.1	−0.8	0	1	Yes
1	0	1	0.2	−0.7	−1.1	−0.9	0	0	No
1	1	0	0.2	−0.7	−1.1	−0.5	0	0	No
1	1	1	0.2	−0.7	−1.1	−1.6	0	0	No
1	0	0	0.2	−0.7	−1.1	0.2	1	1	No

We assume initially randomly generated weights of $W_0 = 0.2$, $W_1 = 0.3$, $W_2 = -0.1$. Clearly this perceptron cannot initially replicate the NOR function, as with inputs X_1 equal to 0 and X_2 equal to 1 the output of the neuron will be 1, which is incorrect. Note the input X_0 is the bias, which is always kept fixed at 1

Table 7.5 Demonstration that the learned weight values correctly compute the NOR function

X_0	X_1	X_2	W_0	W_1	W_2	Activation	Output	OR	NOR
1	0	0	0.2	−0.7	−1.1	0.2	1	0	1
1	0	1	0.2	−0.7	−1.1	−0.9	0	1	0
1	1	0	0.2	−0.7	−1.1	−0.5	0	1	0
1	1	1	0.2	−0.7	−1.1	−1.6	0	1	0

7.8.6   Deep Learning and the Backpropagation Algorithm

While the single-layer perceptron performed well on some pattern recognition tasks, on other apparently simple tasks it did not. For example, a single-layer perceptron can never be taught the exclusive-or function.

Minsky and Papert, in their book Perceptrons (Minsky and Papert 1969), performed a comprehensive investigation of the limitations of the single layer perceptron, demonstrating its shortcomings in some detail. Basically, a single-layer perceptron will learn, and has been proven to converge to a solution for any linearly separable function. However, the single-layer perceptron cannot solve linearly inseparable functions. In order to overcome this problem, we may add another layer or layers) to the structure; however, if we continue to use a step function as the thresholding function, then there is a problem with credit assignment. In order to overcome this issue, we can smooth out the function so that more information is passed through the network. A common smoothing function is the sigmoidal function

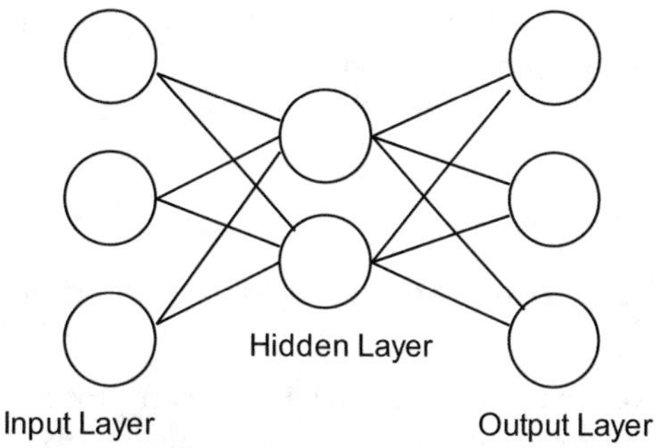

Fig. 7.18 Simple multi-
layer perceptron

Hidden Layer

Input Layer   Output Layer

$$f(\lambda_{network}) = \frac{1}{1 + e^{-(\lambda_{network})}}$$

where higher values of $\lambda_{network}$ cause the function to approach the hard-limiting model.

Now, combining several layers of these units together we construct the multi-layer perceptron which is, indeed, able to solve linearly inseparable problems (see Fig. 7.17). Indeed, a theorem by Kolmogorov (1957) states that any continuous function can be computed using a three-layer perceptron using continuously increasing non-linearities. The theorem does not, however, state how the weights might be chosen (Fig. 7.18).

A training algorithm has however been developed for the multi-layer perceptron and indeed it was the development and popularisation of this algorithm—the backpropagation algorithm in 1986 (Rumelhart et al. 1986) that led more or less directly to the revival of interest in the field in the 1980s which has continued to this day, with another major upsurge of interest happening in the last few years. The algorithm is a gradient search technique and an early demonstration of the power of the algorithm was given by Sejnowski and Rosenberg (1986) in their NETtalk system. They trained a two-layer perceptron with 120 hidden units and over 20,000 weights to learn to successfully translate letters to phonemes. While the backpropagation algorithm has found application in a wide area of domains, issues do however exist. One of the major problems with backpropagation is that, as it is a gradient search technique, a local minimum may be found instead of the required global minimum; also in many cases the number of presentations of the training data may be very large before convergence occurs, if indeed it does occur. Another problem that arises is in the addition of extra learning data after the network has been trained on a particular data set. This may cause corruption of already learned data and indeed may require the network to be trained again from scratch. A final disadvantage of the backpropagation algorithm as applied to training multi-layer perceptrons is that it is inherently a supervised learning algorithm, that is for each pass through the training set the desired or expected output must also be presented.

However, given these limitations, the multi-layer perceptron, using back-propagation or some other differentiation algorithm at its core, coupled with modern high-power massively parallel computing technologies can form highly effective and powerful learning engines and can indeed even be viewed as "the quintessential deep learning models" (Goodfellow et al. 2016).

7.9   Introduction to Reinforcement Learning

Reinforcement learning involves a process of trial and error, with sequences of actions resulting in favourable final outcomes being rewarded, and/or sequences resulting in non-favourable outcomes being punished in some fashion depending on the particular RL variant employed. There are three core components in a system employing reinforcement learning (RL). These are the *policy*, which determines the learning agent's behaviour at any particular point in time, the *reward function*, which broadly speaking provides immediate feedback on the suitability of individual states in the search space, and a *value function*, which provides a more long-term assessment of the desirability of a particular course of action based on the maximisation over time of rewards immediately available, in conjunction with those rewards considered likely to accrue based on this particular course of action.

The core RL problem then reduces to the determination of an effective policy, working in conjunction with a learned value function, which can be successfully applied to the problem domain at hand.

7.9.1   *Reinforcement Learning Example Using Noughts and Crosses*

Let us now illustrate the application of reinforcement learning to the game of noughts and crosses (tic-tac-toe), which we addressed earlier using the minimax procedure. One major disadvantage of minimax is that it does not allow us to adapt in order to compete with imperfect play from our opponent. For example, minimax will never make a move that it calculates will result in certain defeat by its opponent. This is because it assumes that its opponent plays in exactly the same "mirror-fashion" as itself, even though it may have been demonstrated through repeated games that the opponent has a "blind spot" and to make this move would in fact quickly lead to the defeat of the opponent. Reinforcement learning allows us to take advantage of imperfect play by our opponent, and, in fact, even to adapt to gradual changes in our opponent's behaviour.

We will look at an example based on that was introduced by Sutton and Barto (1998) and also discussed in Luger (2009). In our treatment here, we will work through several complete games in order to demonstrate clearly the development of an optimal control policy.

Fig. 7.19 State assigned to
the value 0.6

Our reinforcement learning agent will operate using a table of numbers, one for each possible state in the game. Each of these numbers represents the estimated probability of winning, given that a move is taken that leads us to this state. This whole table of values we then treat as our value function.

Playing as X, and using our RL agent, we assume very little a priori knowledge as to which states are better or worse, apart from the fact that moving to any state with three X's in a row is a guaranteed win and so is assigned value 1; similarly any state with three O's in a row is a guaranteed loss and assigned value 0. For the purposes of this illustration all drawing states are also treated as losses and assigned value 0. All other states are assigned a value of 0.5 initially, except for the state where we place an X in the centre square, to which we assign the value 0.6 (Fig. 7.19).

This is the single piece of *a priori* knowledge that we will allow to our RL algorithm; that is, that playing in the centre square initially may lead to marginally more favourable outcomes overall. All other nodes, apart from winning and losing positions, are assigned a value of 0.5, as we assume no a priori knowledge of the relative worth of these nodes.

7.9.1.1   Fixed-Policy Strategy

Our fixed-policy opponent we will assume follows the following strategy:

Algorithm FIXED-POLICY-MOVE
If winning-move-possible **then**
make-winning-move
else if centre-square-available **then**
play in centre square
else if any-corner-square-available **then**
play in a corner square (at random)
else play in a side square (at random)
End If

This algorithm, while simple, looks on the face of it to be reasonable. If it detects a winning move in the next turn it will play it (note that there is no mechanism for detecting whether the RL agent has a winning move open to it—but, then again, the RL agent also has no explicit mechanism for detecting an imminent win by its opponent, (although this may be learned by the RL agent).

If no winning move is available to the fixed-policy agent, it will play in the centre, if available (this bias is also built indirectly into the RL agent through the 0.6 value assignment), otherwise in a corner, if available, else on a side square. So, in a sense, there is a little more "intelligence" built into this fixed-policy agent (it will play in a corner as opposed to the side, opening up more potential winning avenues as there are three possible winning avenues from a corner, as opposed to two from a side square; the RL agent initially makes no such distinction), however, crucially, the fixed-policy agent does not learn.

7.9.1.2   Temporal Difference Learning Rule

We will update our RL agent's winning probabilities using the equation

$$V(S_n) = V(S_n) + c(V(S_{n+1}) - V(S_n))$$

where S_n is the state before the move is made, S_{n+1} is the state after this move is made, preceded by the opponent's response, and $V(S_n)$ is the updated probability for state S_n. The constant c is a small positive value which we fix at 0.2 in this example. This is termed a *temporal difference* (TD) learning rule, as changes are made based on the difference in values at two distinct points in time, n and $n+1$.

Note that in the following analysis we assume *greedy* play by the RL agent; that is, in making a move it considers all possible moves, and then chooses that move with the highest value: if several moves result in states with the same probability value then one of these moves is chosen at random. In order to explore the search space more fully, at certain intervals we may instead choose a move at random. These moves are called *exploratory* moves, and in this case the move will not then alter our learned value function (table of probabilities).

7.9.1.3   Let's Play a Game

Ignoring symmetrically equivalent positions, three moves are initially possible for the RL agent. It will play to the centre square, (a greedy move using the 0.6 value assigned to this state as opposed to a 0.5 value for the other two options) (Fig. 7.20).

The fixed-policy (FP) player will then play to one of the corner squares (as the centre square is taken), leading to the position depicted in Fig. 7.21.

Fig. 7.20 The RL agent's three possible first moves (ignoring symmetries)—estimated values for each state are given in the shaded circles

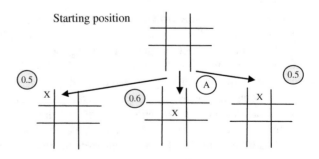

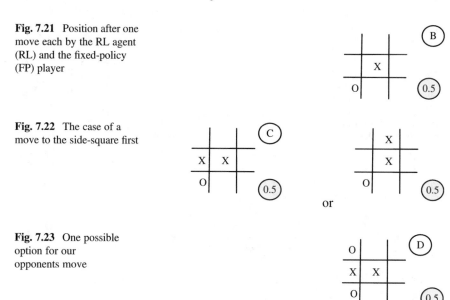

Fig. 7.21 Position after one move each by the RL agent (RL) and the fixed-policy (FP) player

Fig. 7.22 The case of a move to the side-square first

or

Fig. 7.23 One possible option for our opponents move

Ignoring symmetrically equivalent positions, two situations now arise. Either RL moves to a side square, with a guaranteed win for X (as O will now play in a corner), or else the RL agent moves to one of the free corner squares, with mixed options.

Taking the case of a move to a side square first we have the situation as depicted in Fig. 7.22.

Both of these cases lead to the same outcome, with similar analysis, so we will just deal here with the first case, and leave the second as an exercise to the interested reader. We will now generate a backup from this board position to update the value associated with RL's previous board position

$$V(A) = V(A) + c(V(C) - V(A))$$
$$= 0.6 + 0.2(0.5 - 0.6) = 0.58$$

The value for position A in the lookup table is now updated to reflect this value.

Moving on from this position (C) it is now our opponent's turn to move. Because of the fixed policy of moving to a corner square (assuming the centre square is taken), just three options present themselves. All three of these options result in a win for X.

For example, we could have as in Fig. 7.23.followed by our move as in Fig. 7.24.

This move guarantees us to win. Given greedy play on our part we are certain to take this move as it has the highest assigned value (1.0), and, of course, also happens to be a win for us.

Backing up these values, we then get

Fig. 7.24 Our response to
the move in Fig. 7.23

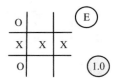

Fig. 7.25 Simultaneous
update of equivalent states

Fig. 7.26 States A and C
showing altered state
probabilities

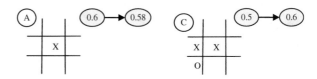

$$V(C) = V(C) + c(V(E) - V(C))$$
$$= 0.5 + 0.2(1.0 - 0.5) = 0.6$$

We may also assume that all equivalent states in the table receive this correct
updated value: e.g. all of the states in Fig. 7.25 receive this same value.

And all go from a value of 0.5–0.6 (together with the other 5 equivalent states).
So, state C is updated to reflect this changed value, and looking at our table we have
now altered the probability value for two states, together with their symmetrical
equivalents, as shown in Fig. 7.26.

7.9.1.4 Let's Play Another Game!

Starting off we again choose state A as it has the highest assigned value (this time
0.58 as opposed to 0.5 for all the other states). Our opponent will move to a corner
square as before. At this point we will now generally choose the same move in
response as before (or an equivalent move if our opponent moves to one of the other
corner squares—remember we have already updated all of the equivalent states in
our table. Assuming our opponent moves to the same square as before we have an
update now of the value of A (based on the value of C)

$$V(A) = V(A) + c(V(C) - V(A))$$
$$= 0.58 + 0.2(0.6 - 0.58) = 0.584$$

marginally increasing the value of state A. Playing as before we win again; this time
we generate an update for node C as before:

$$V(C) = V(C) + 0.2(V(E) - V(C))$$
$$= 0.6 + 0.2(1.0 - 0.6) = 0.68$$

It is clear that in subsequent games X will continue to win, thus reinforcing his winning strategy each time a game is played.

7.9.1.5   Of Course, the Devil Is in the Detail

On the other hand, of course, for her second move X could have played in a corner adjacent to O's move instead of on the side. Two options arise; in each case, firstly the value for state A is updated as before, changing it from a value of 0.6 to 0.58. Then her opponent may play in the opposite corner to X's winning line, as in Fig. 7.27.

In this case X obviously wins in the next move, with similar analysis to the previous case.

The other option is that O plays to block X (Fig. 7.28).

This is a similar position to that which would arise if X had played initially in the opposite diagonal corner to O, followed by O's response in one of the other corner squares, as in Fig. 7.29. so that the analysis that follows covers this situation also.

In this case X has only one move that will avoid a loss. This is because X is moving at random at this point as all states apart from "X in the middle" and the winning states have equal probability, and nought will always play a winning move if this is available.

If X does indeed block O (by playing to the bottom centre square), she will go on to win—the analysis here is quite straightforward, as shown in Fig. 7.30.

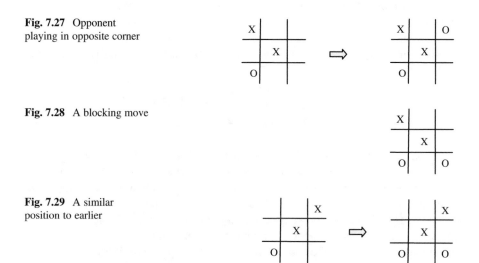

Fig. 7.27 Opponent playing in opposite corner

Fig. 7.28 A blocking move

Fig. 7.29 A similar position to earlier

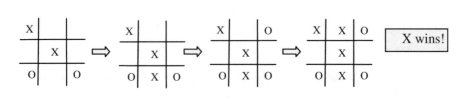

Fig. 7.30 A winning game sequence

Fig. 7.31 An alternative
move for X

Fig. 7.32 A win for O!

Fig. 7.33 Our updated
value for state F

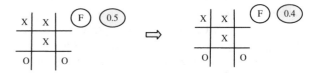

And again, the probability of making this blocking move will be increased from
0.5 to 0.6 for subsequent games. We now look at the case where X does not, in fact,
make this move, and the backups generated. Let us say instead we move as in
Fig. 7.31.

This move will be followed inevitably by a win for O as shown in Fig. 7.32.

It is our move next, however as we have lost, we now instead back up the value
0.0 to state F (Fig. 7.33). We have

$$V(F) = V(F) + c(V(G) - V(F))$$
$$= 0.5 + 0.2(0.0 - 0.5) = 0.4$$

Now, in the next game played X still has a higher probability of playing to the
side than to a corner (three corners as opposed to four sides) but if X *does* play the
corner square for her second move, he is *less* likely to choose this losing move
subsequently (top centre square) because of the lower value assigned to it.

In this fashion we clearly see that, in every circumstance, the RL agent gradually
converges on a winning policy for playing against this fixed-policy opponent,
illustrating the power of the reinforcement learning approach to learning.

7.10   Evolutionary Learning

The term "learning" is not used directly so much with regard to evolutionary algorithms and evolutionary computation in general, perhaps because of the supposed time scales between the parent genome and its offspring. And, of course, in nature this may well be the case; in the case of humans, where this time scale is in the decades (in terms of the inter-generational gap), this is certainly the case. It is clear that there is a large time-scale difference between individual generations (generally several decades) as opposed to, for example, learning not to put one's hand on a hot stove (virtually instantaneous), or learning a language, or learning to walk (months or years in these cases). But if we view learning as adaption and change in a natural or an artificial organism over time in order to produce better "results", then evolution is certainly a learning process—and, of course, with modern computational power we have the means to telescope decades or years of evolution down into days, hours, or even seconds or milliseconds.

7.10.1   Genetic Algorithms

We introduced initially the broad field of evolutionary algorithms in Chap. 5, here we will give a little more detail on the workings of one of the most commonly employed evolutionary algorithms—the genetic algorithm. To illustrate the process of evolutionary (phylogenetic) learning and adaptation we will demonstrate the evolution of ball-kicking behaviour in a humanoid robot—obviously a key element in a soccer-playing robot capable of competing in the RoboCup robot soccer tournament. As this example falls under the general category of *Evolutionary Humanoid Robotics* (EHR) we will leave the details of this example until Chap. 8. Our introduction to genetic algorithms here is based on Chap. 2 of the book with the same title (Eaton 2015), to which the interested reader is referred to for further detail.

Genetic algorithms, as originally proposed by Holland (1975) employ three basic operators: reproduction, crossover, and mutation. Starting off with an initial population of structures, typically although not always, chosen at random, subsequent populations are generated by the application of these three operators in conjunction with a fitness function, which is simply a measure of the performance of an individual at the task in hand.

The most basic formulation of genetic algorithms involves populations of bit strings of fixed length, and this is the configuration we will introduce here. Typically, starting from the initial population individuals are selected based on their fitness relative to other individuals in the current population. Several possibilities then exist—either the individual is reproduced and copied directly into the next generation, or else is altered before copying using either the mutation operator, the crossover operator, or both. This process is continued for subsequent generations, until either a fixed generation limit is reached, or a satisfactorily level of performance

| x1 | x2 | x3 | x4 | x5 | x6 |

X=

| y1 | y2 | y3 | y4 | y5 | y6 |

Y=

Fig. 7.34 Two simple bit strings prior to crossover

X'=

| x1 | x2 | y3 | y4 | y5 | y6 |

Y'=

| y1 | y2 | x3 | x4 | x5 | x6 |

Fig. 7.35 Bit strings after crossover

is reached. The change in performance levels over generations is typically tracked using graphs of the maximum and average performance levels of populations over the generations.

For bit strings the mutation operator simply involves flipping a bit from a 1 to a 0, or vice versa. In genetic algorithms the probability of mutation is typically kept quite low compared to the probability of crossover.

The crossover operator is generally seen as the key to the successful operation of genetic algorithms. We will look at a simple example of crossover in operation. Two individuals X and Y are selected for crossover, each, say of length 6 bits, as shown in Fig. 7.34.

The next thing to do is to select a crossover point, for more complex crossover operations several points may be chosen, but the principle is still the same as for the single point crossover operator illustrated here. This crossover point is chosen at random in the interval $1 \ldots l-1$, for a bit string of length l. If we assume a crossover point of 2 in our example, the strings after crossover would look as in Fig. 7.35.

So, by using this crossover operator over time, which allows us to potentially combine the best bits of two parents into a single individual in combination with the mutation operator together with a carefully designed fitness function, we would hope to see an increasing population and individual fitness as the evolutionary process progresses.

Chapter 8
Computers, People, and Thought

> Looking (and dreaming) toward the future, one can imagine
> nano-scale (bioware) systems becoming a reality, which will
> be endowed with evolutionary, reproductive, regenerative,
> and learning capabilities. Such systems could give rise to
> novel species which will coexist alongside carbon-based
> organisms.
>
> —Moshe Sipper et al. (1997) A phylogenetic, ontogenetic,
> and epigenetic view of bio-inspired hardware systems. *IEEE
> Transactions on Evolutionary Computation, 1(1), 83–97.*

Sipper et al.'s farsighted article quoted from above envisioned a grand scheme where
a variety of techniques including artificial neural networks, cellular automata, and
evolutionary algorithms might be used in the future to create "novel species which
will co-exist alongside carbon-based organisms". No such entities exist today,
however as a first step, in this chapter, we will look at the application of one of
these techniques, arguably the most powerful—evolutionary computation—in the
creation of novel behaviours in a real humanoid robot.

8.1 Evolutionary Robotics (ER)

Evolutionary robotics (ER) involves the application of evolutionary techniques to the
generation of either the "brain" (control systems) or to the "body" (morphology) of autonomous robots, or perhaps both.

So began the chapter on evolutionary robotics in my recent book *Evolutionary
Humanoid Robotics* (Eaton 2015), on which this section is broadly based. Central to
the operation of ER (and, indeed, evolutionary algorithms in general) is the creation
of a so-called fitness function; another core issue with ER in particular is the reality-
gap issue. We will discuss both of these topics briefly shortly.

Firstly, however there is the question of what, exactly, it is that we are using our
evolutionary techniques to evolve. As is clear from the above quote we may use
evolutionary techniques to evolve either the control systems of the robot, or the

© Springer Nature Switzerland AG 2020
M. Eaton, *Computers, People, and Thought*,
https://doi.org/10.1007/978-3-030-55300-5_8

morphology of the robot, or, in some cases both the morphology and control systems. Typically, if the morphology of the robot is being evolved, we also evolve the control systems to a greater or lesser extent. In Karl Sims' early ground-breaking 1994 article "Evolving Virtual Creatures" (Sims 1994a) he demonstrated in simulation the evolution of "creatures" composed of interconnected articulated three-dimensional rigid components. Both the structure of each creature, each of which was represented by a directed graph of nodes and their connections, and the control mechanism, which was implemented as a neural structure were subject to the evolutionary process. A variety of behaviours were evolved including walking, jumping and swimming. Sims made the following far-sighted observation on these early experiments (Sims 1994b), and as quoted in Eaton (2015)

> As computers become more powerful, the creation of virtual actors, whether animal, human, or completely unearthly, may be limited mainly by our ability to design them, rather than our ability to satisfy their computational requirements. A control system that someday actually generates "intelligent" behaviour might tend to be a complex mess beyond our understanding.

Sims' work, while far-seeing and influential, just involved the evolution of simulated creatures, not embodied robots capable of interactions in the real world. A number of researchers would question the utility of such an approach (Flynn and Brooks 1989)

> One real robot is worth a thousand simulated robots. Building robots that are situated in the world crystalizes the hard issues.

Much of the early work in evolving behaviours on real (embodied) robots involved evolving a variety of behaviours for wheeled robots. An early classic experiment of this nature was that by Floreano and Mondada in 1996 and described in their article "Evolution of homing navigation in a real mobile robot" (Floreano and Mondada 1996). This describes a series of evolutionary experiments carried out entirely on a real mobile robot (no simulations performed) without human intervention for the evolution of homing behaviour on a small wheeled mobile robot. The robot used for the experiments was the Khephera miniature mobile robot, which has been widely used in early evolutionary robotics experiments. Details of these, and many other evolutionary robotics experiments are given in Nolfi and Floreano's book *Evolutionary Robotics* (2000), which still remains a standard reference work for evolutionary robotics experimenters. Also see Bongard (2013) for a concise introduction to the field.

8.1.1 From the Reality Gap: Into the Uncanny Valley?

However, in most cases of evolutionary robotics experimentation a simulator is utilised for two main reasons. Firstly, the strain on the robot's actuators caused by repeated experimentation over a long time scale is a major issue—this strain is accentuated if we move from simple wheeled robots to more complex many degrees-of-freedom robots such as humanoid robots, the main focus of this chapter. The other main issue is that of time—in simulation we can telescope experiments that might take weeks or months on a real embodied robot into hours or days. The final simulated evolved robot body/brain can then be transferred onto the real robot for further testing. However, a problem arises here—what if the simulator does not accurately model the behaviour of the real robot? This is the so-called *reality gap* issue which may result in inefficient, incorrect, or even dangerous evolved behaviours. Several approaches have been taken to close this gap; space precludes further discussion of these measures, however for a further concise discussion of this topic see Eaton (2015).

Another interesting effect, relating more to specifically to humanoid robot evolution or design relates to the so-called *uncanny valley* (Mori 1970). Let us say the we have managed to evolve (or otherwise design) a robot which is while not exactly lifelike in terms of similarity (looks and behaviour) to a human, but very close. This gives rise to a strange effect which can cause unease in a human observer. For a robot which is in no way similar to a human there is no issue, as there is no question of mistaking that robot for a human. Similarly, for a robot which is identical in every way to a human again there is no issue, as we assume the robot *is* human. However, in between these two extremes, and especially as we closely approach human likeness and behaviour lies the uncanny valley where a sense of eeriness grips the average human observer. Incidentally his fact is used to good effect by the makers of zombie films.

8.1.2   Fitness Function Design

In evolutionary robotics the fitness function should specify *what* task it is that we want the robot to perform and not *how* it is to perform this task (Eaton 2015). It is fair to say that fitness function design is one of the most critical tasks for the evolutionary robotics practitioner. In the words of Nolfi and Floreano (2000)

> The more detailed and constrained a fitness function is the closer artificial evolution becomes to a supervised learning technique, and less space is left to emergence and autonomy of the evolving system

So a delicate balance needs to be struck in the formulation of the fitness function for a particular robot application, because of this the design of a fitness function tends to involve a number of iterations, with the experimenter at each stage tweaking the function a little in order to bring the observed robot behaviour closer to the desired outcome. This process is discussed in further detail in Eaton (2015). Space precludes a further detailed discussion of this interesting topic, however for a detailed discussion of fitness function taxonomy and design see the excellent paper by Nelson et al. (2009).

8.2   Evolutionary Humanoid Robotics (EHR)

Evolutionary humanoid robotics (EHR) involves using artificial evolutionary techniques to evolve all or part of a humanoid robot's body and/or brain. It has its roots
back in the early 1990s with Hugo de Garis's work on the evolution of bipedal
locomotion in simulation for a pair of stick legs (de Garis 1990a–c), and Karl Sims'
work on the evolution of both the body and brain three dimensional simulated
"creatures" mentioned earlier, and which could take a humanoid form. Both de
Garis's and Sims' work were purely in simulation with no application to a real
humanoid robot. One of the earliest applications of evolutionary techniques applied
to a real robot was Arakawa and Fukuda's (1996) application of a genetic algorithm
to determine joint angles for a real humanoid robot made of aluminium, with
13 joints. The evolution was conducted in simulation and then transferred to the
real robot resulting in successful walking behaviour.

Since these early experiments much research has been conducted on the application of evolutionary techniques to both the generation of a variety of behaviours for
humanoid robots, and, less frequently to the evolution of the gross anatomical
structure of the robot. Evolved behaviours include bipedal locomotion (walking),
dance, ball kicking, crawling, jumping, ladder climbing, and grasping and manipulation. Much of this research has been conducted in simulation only, with no transfer
to a real robot, however in many cases the evolved behaviours have been transferred
successfully onto a real robot platform, such as the Nao humanoid, as we will
illustrate in the next section.

8.2.1   Application of EAs to Ball-Kicking Behaviour

In keeping with our focus on game playing applications and in the spirit of the
RoboCup grand challenge introduced earlier we will now look at the application of
evolutionary algorithms in the evolution of stable and robust ball-kicking behaviour
for a humanoid robot. The humanoid robot in question is the Nao robot, which we
introduced earlier in Chap. 5 in our general discussion of humanoid robot platforms.
Nao is particularly interesting in this context as it is the robot used in the RoboCup
Standard Platform League (SPL) competition held annually as part of the RoboCup
series of robot soccer competitions.

Our general procedure will be as follows. Using a fitness function based mainly
on the distance travelled by the ball following the kicking motion by the robot using
a fixed time frame, in conjunction with the length of time that the robot remains in an
upright position following the kick, we aim to evolve over succeeding generations
ever fitter ball-kicking abilities in our Nao robot.

Initially evolution takes place in simulation; this eliminates the possibility of
damage to the Nao humanoid or to its environment, an especially important factor in
the early days of the evolutionary process, when the robot is just "finding his feet",

so to speak. Evolution in simulation also obviates the need for power tethers to the robot, and thus simplifies the whole mechanics of the evolutionary process. Once we have evolved stable robust kicking behaviours in simulation, we then need to transfer these evolved kicks onto the real Nao robot for further evaluation and perhaps subsequent deployment in a "real-world" robot soccer competitive environment.

8.2.1.1 Experimental Details

The fitness function for these experiments was based on the distance the ball travels within a certain time frame assuming that the robot remains upright and doesn't fall over within the time frame (set at 5 s for all of the experiments). If the robot fails to move the ball but remains upright over the course of the time frame it is given a fixed reward, this reward is reduced proportionally based on the length of the time the robot remains upright. Proportionate reward is also given in the case where the robot moves the ball some distance, but subsequently topples over.

The bit string representing each individual is 416 bits in length. This allows for the encoding of the joint angles for each of the 24 joints of the Nao robot in 4-bit chunks for each of four keyframe values. Two further 16-bit values encode the maximum and minimum ranges for each of the joints, and the movement durations for each of the four keyframes. The Nao robot then cycles through the joint values contained for each of the four keyframes until the time limit is reached, or it falls over. To avoid jerking motion between the individual keyframes an interpolation function is used to smooth out the joint movements. The interested reader is referred to Eaton (2016) for further experimental details.

8.2.1.2 Tackling the Reality Gap: The Dual-Simulator Approach

Unfortunately, because of subtle (and sometimes not so subtle) inaccuracies in the simulation environment—the so-called "reality gap" issue, where behaviours evolved or otherwise generated in simulation do not exactly correspond to these same behaviours in reality—there is still a significant risk of unstable and perhaps even damaging behaviours being evinced on transfer to the real humanoid robot.

To minimise this possibility we introduced a second simulator into the equation— the *dual-simulator* approach, where behaviours evolved on the first simulator are first transferred to the second simulator for initial testing, and only when successful stable kicks are observed on both simulators are these kicks finally transferred onto the real Nao robot. Using this dual-simulator approach we find that a high percentage of evolved kicks (over 90% on initial experimentation) transfer accurately and stably to the real humanoid.

Put simply, this approach relies on the observation that different glitches or inaccuracies are likely to occur in the two simulation environments, but the likelihood of exactly the same inaccuracies occurring in each distinct simulation

Fig. 8.1 An example of an evolved kick, as transferred from V-REP to the Webots simulator; The two small boxes in the top and bottom left of each of the four frames give the scene as perceived by the Nao humanoid for each of the frames, from the top and bottom cameras respectively of the Nao robot. The top camera, mounted on the robot's forehead, focuses straight ahead, whereas the bottom camera, located in the chin area, focuses on the feet of the robot. Read from *top left* to *bottom right*

environment is low; hence if stable behaviour is observed in both simulation environments there is a high probability of successful transfer across the reality gap onto the real humanoid.

Figure 8.1 shows an example of a successful kick, as evolved in the V-REP simulator and subsequently transferred to the Webots simulation environment. Figure 8.2 then demonstrates this kick as implemented on the real Nao humanoid robot, resulting in a stable whole-body kicking motion. Maximum and average figures for the evolutionary process as averaged over three runs for 500 generations are given in Fig. 8.3. In these graphs fitness values of over 2500 generally correspond to a stable kicking pattern.

Fig. 8.2 The evolved kick from Fig. 8.1 as finally implemented on the real Nao robot. Compare these whole-body motions with the simulated kick from Fig. 8.1. Read from *top left* to *bottom right*

8.3   Where to from Here?

So—what exactly have we demonstrated here? Clearly by no stretch of the imagination could a small humanoid robot with the (albeit evolved) ability to kick a ball (however effectively) be interpreted as being imbued with human level thinking and computational capabilities. However, harking back to the quote at the start of this

Fig. 8.3 Maximum and
average fitness, averaged
over three runs for
500 generations for the
evolution of ball-kicking
behaviour

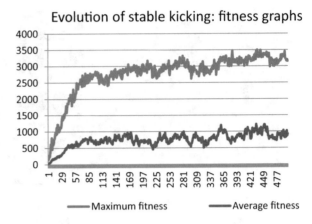

chapter from Moshe Sipper we do have here a certain "proof of concept" that
perhaps in the distant (or not so distant) future artificial analogues of the evolutionary
process that is thought to have been the major developmental force in the development of life on earth—up to and including the development of intelligent conscious
creatures including man—might indeed be harnessed in the creation of intelligent
humanoid robots. These robots might in turn, someday, be accepted as separate
conscious entities in their own right, with their own set of rights and responsibilities.

And this is indeed a profound possibility—one we will now address in more
detail in the final part of the book where we look to the wider picture and the various
issues that may arise in the future (indeed some of which are with us today) in the
assimilation of these potential advanced technologies in society. To conclude this
chapter, I quote from p.31 of *Evolutionary Humanoid Robotics* (Eaton 2015)

> ...it seems clear that evolutionary techniques will play a significant role in the development
> of future humanoid robots. Exactly the scale of that contribution is not yet clear; however,
> there exists a clear proof of concept in the overarching power of the natural evolutionary
> process: humanity itself.

Part III
The Wider Picture

Chapter 9
Ethical, Societal, Philosophical, and Spiritual Issues

9.1 Introduction

The history of life on this earth that is recorded in the rocks is full of evidence of races of living things that have populated the earth for a long time and then become extinct, such as the dinosaurs. In that long history, rarely does a race survive... There seems to be no kind of escape possible. It is necessary to grapple with the problem: How can we be safe against the threat of physical harm from robot machines?

...

The robot machine raises the two questions that hang like swords over a great many of us these days. The first one is for any employee: What shall I do when a robot machine renders worthless all the skill I have spent years in developing? The second question is for any businessman: How shall I sell what I make if half the people to whom I sell lose their jobs to robot machines?
—Edmund Berkeley, *Giant Brains, or Machines that Think*

The above two quotes elucidate two of the greatest fears extant today about the seemingly imminent arrival of highly capable A^2IEs. Firstly, the possible development of a potentially hostile superintelligence intent on the extermination of humanity, or the relegation of humans to the status of mere pets. Secondly, the possibly catastrophic loss of jobs to our new artificially intelligent companions, as evidenced most clearly nowadays by the issue of self-driving cars and trucks.

But—as is clear from some of the language used in the quotes—these words were not written last week, or last year, or even in the last few decades. They were, in fact, written 70 years ago by the eminent Edmund Berkeley, one of the seminal figures in the early history of computing, whose bestselling book *Giant Brains* brought computing to the masses in 1949. So—we've seen this all before, haven't we? Well—what's different this time around? *Is* there anything different this time around? I would argue *yes*.

In Part III we will look at potential future scenarios in the creation and application of AIEs/AI^2Es. Apart from the two important topics mentioned above, another area of key concern to many researchers is that of the potential development of lethal

© Springer Nature Switzerland AG 2020
M. Eaton, *Computers, People, and Thought*,
https://doi.org/10.1007/978-3-030-55300-5_9

autonomous weapons systems: that is weapons capable of applying lethal force without any necessity for human intervention. We will further address this important topic in Chap. 12. We will also address potential threats related to individual and collective privacy and associated issues related to freedom and democracy in Chap. 13. Then in the final chapter we will look at the future of society—the choices and decisions we make now and into the near future, and their potential long-term consequences.

In this introductory chapter we will address the notion of "technology for technology's sake", some potential implications of an ever more connected world spurred on by advances in computer and communications technologies, and we will also broach the controversial topic of religion and spirituality in relation to advanced artificially intelligent entities.

In a sense maybe you could say that, at the minute, we are collecting the parts of a great jigsaw puzzle; each part has a particular shape, and a certain portion of the overall picture. We are fairly confident at the minute that we have an idea of what the whole jigsaw should look like, but uncertain as to what the individual parts of the jigsaw should be exactly. But once the parts have been created it should not be the most arduous task to connect them together to create the whole. And, if there are seen to be gaps in the picture, it should be fairly clear what these gaps are, and concerted efforts can be made to fill them.

We (AI/IS/AIE/Robotics researchers) are currently at the stage of having found/ created a substantial number of the jigsaw puzzle pieces. When a sufficient number have been found, it may well start to become clearer how they fit together to make the whole jigsaw. And then, as any avid jigsaw puzzle enthusiast will tell you, all will change.

9.2 A Machine-Centred Orientation Towards Life?

Society has unwittingly fallen into a machine-centered orientation to life, one that emphasises the needs of technology over those of people....

Worse, the machine-centered viewpoint compares people to machines and finds us wanting....

So begins the prophetical text *Things that make us smart: defending human attributes in the age of the machine* by Donald Norman. What is of interest here is that this book was published in 1993, in the years preceding the uptake by the masses of the now ubiquitous Internet, which facilitates the World Wide Web, social media, and associated technologies. In fact, what Donald Norman was particularly concerned with at the time was television, now generally regarded as a relatively benign technological force compared to the Internet and WWW.

9.2.1 Advanced Technology for Good: Or Just Technology for Technology's Sake?

We can take, as an example, another "advancement" that has come up for some scrutiny in recent times—electronic voting machines. Again, do the undoubted advantages that such devices offer (very rapid access to voting results, etc.), outweigh the significant concerns expressed by many about the potential for fraudulent results, potentially undermining the whole basis for the western democratic system? In Ireland we answered this question with a resounding "no" a couple of years ago, with the resulting destruction and sale of several million euros worth of new-fangled electronic voting devices. Unfortunately, these devices lacked one simple component—the ability to verify that what they presented as the votes cast for particular candidates or parties was, in fact, correct.

As another example of an "advanced technology" we might take the cordless rechargeable clothes iron. On the face of it this is a useful (and expensive) gadget. It is light, and it does away with the need for a cord, which can be a major nuisance. My wife purchased such a device a little while ago, and it certainly looks the business. However—there is one small problem. When in use it requires replacement on a special charging stand every 10 s or so for a short period (a couple of seconds) to recharge. When ironing a shirt or two this is not a major issue, but over any extended period this is a major inconvenience. "Advanced" technology? Maybe. But a useful innovation as it currently stands? Not really.

9.2.2 Just Because We Can...

Bill Clinton, the former US president, asked about his relationship with the former White House intern Monica Lewinsky, is once famously reported as saying

> I think I did something for the worst possible reason—just because I could.

Would it not be a damning indictment of early twenty-first century society regarding our use and promotion of certain clearly damaging (to humanity as a whole) technologies to say, "we did it just because we could"?

A similar sentiment was expressed by panellists in the recent (May 1, 2018) "Techethics Conversation" hosted by the IEEE (Institute of Electrical and Electronic Engineers) with the theme of "Ethical Considerations 200 Years After Frankenstein". This can be summarised as: given that it is hard to predict what certain technologies will result in, just because we *can*—does it mean that we *should*?

9.3   Humans: Not Their Machines—The Imminent Threat?

What many would consider to be an even more worrying issue at the present time, and certainly into the near-term future, are the potential human (ab)uses of advanced technological tools, rather than humans being dominated by machines of their own creation. That is not to say that the latter prospect is an impossibility, rather that it is a far more distant prospect at present.

Facilitating many of these developments is the extraordinary behemoth that we-humankind has created—the worldwide Internet, And I do say "we created" because it appears to be a common misconception, especially to those under the age of 30, that the Internet in all of its cultural manifestations—the World-Wide-Web, Facebook, Instagram, etc.—has always existed, and that these technological innovations are, in a sense, some sort of divine plan. This, of course, is not the case. The Internet was developed by us, humanity, and it is within our remit to make changes to its operation, or (highly unlikely!) to completely eliminate it in its present form.

Surprisingly, perhaps the most serious imminent threat to humanity from technological (ab)use may not be from superintelligence, killer robots, or even jobs issues It may well be from that seemingly innocuous device that a high proportion of the human population now carry around with them every day (over 90% of Irish adults own one, and the numbers are rising), and have beside them on the bedroom locker at night—the smartphone.

But not just the smartphone per-se, but one particular set of applications that run on this device—the so-called "social media" applications. Of course you can access social media from other platforms also, such as from desktop computers, laptops, etc., but none bear the immediacy of the all-ubiquitous smartphone. Let us consider an oft-referenced quote from Sean Parker,[1] the first president of Facebook and often credited as the person who took Facebook from the university campus into the multi-billion enterprise it is today, talking about his creation:

> "... it literally changes your relationship with society, with each other ... It probably interferes with productivity in weird ways. God only knows what it's doing to our children's brains." "The thought process that went into building these applications, Facebook being the first of them, ... was all about: 'How do we consume as much of your time and conscious attention as possible?'" "And that means that we need to sort of give you a little dopamine hit every once in a while, because someone liked or commented on a photo or a post or whatever. And that's going to get you to contribute more content, and that's going to get you ... more likes and comments." "It's a social-validation feedback loop ... exactly the kind of thing that a hacker like myself would come up with, because you're exploiting a vulnerability in human psychology. "The inventors, creators — it's me, it's Mark [Zuckerberg], it's Kevin Systrom on Instagram, it's all of these people — understood this consciously. And we did it anyway."

In the words of Max Tegmark "if we don't know what we want we're unlikely to get it" (Tegmark 2017). Of course, we can turn this phrase a little on its head and say, "be careful what you wish for—because you might just get it". For example, do we

---

[1] https://www.axios.com/sean-parker-unloads-on-facebook-god-only-knows-what-its-doing-to-our-childrens-brains-1513306792-f855e7b4-4e99-4d60-8d51-2775559c2671.html

really wish for ever more powerful smartphones and other devices yet to be dreamt of in the hands of an ever younger cohort of users—12, 13 year olds (an average daily usage of approx. 190 min for 12–17 year olds), and even 9 and 10 year olds? In fact, a recent survey of phone usage by teenagers and children in Germany showed that 38% of children aged between eight and nine owned a mobile phone, of which 18% were fully internet-enabled smartphones. The future—for now—is in our hands; let us choose wisely!

And this area, scary as it is, ties nicely into the theme of this book—computer algorithms which are now being used to modify people's thought processes. Computers, people, and thought; but not a benign combination by any stretch of the imagination!

9.4 Outline Rating of Potential Threats

We may choose to rate the potential threat posed by advanced technologies in terms of the two broad categories:

a) *Imminence* and
b) *Seriousness of potential threat.*

Looking at a possible gradation of these two categories we might have:

- Imminence

    0. Now
    1. <10 years
    2. 10—30 years
    3. >30 years

- Seriousness of potential threat

    0. Complete destruction of humanity
    1. Serious consequences for a large section of humanity
    2. Serious consequences for a small section of humanity/relatively minor consequences for a large section of humanity
    3. No major adverse implications for humankind

So, based on this outline classification a level 3-0 threat constitutes a technological advancement with the potential for the complete annihilation of humankind, however with a timescale of over 30 years (which should allow some time to analyse and potentially defuse this threat). Table 9.1 lists some possible issues that may arise from the development of AIEs at present and in the future.

Some analysts make predictions based on timescales of 50 years or more. Given the current scale of technological advancements I would consider it virtually impossible to make predictions with any degree of credibility this far into the future.

Table 9.1 Some issues potentially arising from the development of AIEs, and a tentative time scale relating to their potential future effects

Timescale/issue	Right now	Near future	Medium/long term	Maybe never
Privacy issues	✓	✓	✓	
Killer robots/autonomous weapons	✓	✓	✓	
Jobs concerns	✓	✓	✓	
Humans as pets			✓	✓
Concerning robot "rights"		✓	✓	
Potential technological "glitches"	✓	✓	✓	
Threats to the democratic process	✓	✓	✓	
Robot "priests"	✓	✓	✓	

However, we would do well to bear in mind "Amara's law", as quoted by Dorn (2015).

> We tend to overestimate the effect of a technology in the short run and underestimate the effect in the long run.

Finally, let us not also forget the enormous potential for good that these new and advanced technologies could have for all of humankind, if we are prepared and determined to circumnavigate the many serious potential pitfalls (some of which might well lead to the destruction of humanity itself).

9.4.1 AI Safety Research

To give some idea of the prevalence of current work in the area of AI Safety, at the time of writing a total of 16 separate bodies are in receipt of grants from the Open Philanthropy Project (www.openphilanthropy.org). Seven of these grants are for sums in excess of US$1,000,000, the highest award (US$5,555,550) being to the UC Berkeley Center for Human-Compatible AI, led by Stuart Russell. The top five grants are listed in Table 9.2, together with their focus areas and the amount awarded.

The Future of Life Institute (FLI) also awards grants for AI safety, the most notable of these being for a sum of US$1,500,000 to Nick Bostrum of the University of Oxford, author of the book *Superintelligence: Paths, Dangers, Strategies*, which propelled the issue of AI Safety into the forefront of many peoples' minds, giving serious credibility to the issue.

Table 9.2 Top five grants in the area of *Potential Risks from Advanced Artificial Intelligence* as awarded by the Open Philanthropy Project

Grant	Organization name	Focus area	Amount
UC Berkeley—Center for Human-Compatible AI	UC Berkeley	Potential risks from advanced artificial intelligence	US $5,555,550
Machine Intelligence Research Institute—General Support (2017)	Machine Intelligence Research Institute	Potential risks from advanced artificial intelligence	US $3,750,000
Montreal Institute for Learning Algorithms—AI Safety Research	Montreal Institute for Learning Algorithms	Potential risks from advanced artificial intelligence	US $2,400,000
Future of Humanity Institute—General Support	Future of Humanity Institute	Potential risks from advanced artificial intelligence	US $1,995,425
UCLA School of Law—AI Governance	UCLA School of Law	Potential risks from advanced artificial intelligence	US $1,536,222

Source: www.openphilanthropy.org

9.5 Spirituality and Religion in the Age of the Robots

...the earth was without form, and void; and darkness was upon the face of the deep. And the Spirit of God moved upon the face of the waters.
  —Genesis, Chapter 1

...the social dangers of our new technology and the social obligations of those responsible for management to see that the new modalities are used for the benefit of man, for increasing his leisure and enriching his spiritual life, rather than merely for profits and the worship of the machine as a new brazen calf.
  —Norbert Wiener (1954). *The Human Use of Human Beings*. Houghton Mifflin Company, New York.

Just as in my previous book *Evolutionary Humanoid Robotics,* where I hesitated long and hard about the inclusion of a chapter on social and ethical aspects of advancing humanoid robot technology and other potentially highly world-altering technologies, so in this text I also hesitated to include a section on the spiritual and religious aspects of current and future AI technologies. In a sense it is the elephant in the room, and this will, for many, undoubtedly be the most controversial section of this book. To be clear, the purpose of this section is not to make the reader question in any sense their religious beliefs (or lack thereof), but to examine what role, if any, religion and spirituality might play in the development and deployment of future A^2IEs. There are many who will argue that there is no place in a text such as this for this type of discussion. And you may well be right. If this is your belief, I suggest you skip over this short section. But I feel it would be remiss of me not to at least touch on these topics.

Some scientists/researchers in the field are quite clear about their religious beliefs. However, owing to the private nature of the subject the majority choose not to highlight their religious beliefs (if any). This is entirely understandable in the current intellectual and social climate where, certainly in some intellectual circles, one is considered something of an oddity, if not an immediate subject of ridicule, for expressing any formal religious beliefs.

The eminent roboticist, Ronald Arkin, is one of the exceptions to this regime. In the introduction to his highly acclaimed book *Behaviour Based Robotics* (Arkin 1998) book he affirms his religious affiliations quite clearly. Equally many researchers in this general field give free rein to their atheistic outlook on life; Richard Dawkins is probably an exceptional example in this regard. Dawkins conducted some of the first of what would now be called "Interactive Evolutionary Computation" experiments, and described the results of his experimentations in his book "The Blind Watchmaker"(Dawkins 1986) in which he discounted the possibility of the existence of God based partly on his observations using these experiments.

9.5.1 Science Versus Religion: Why the Battle?

> God and evolution are not mutually exclusive. Evolution is a scientific phenomenon, one that scientists can study because it is observable and predictable. But digging up fossils does not disprove the existence of God or a higher purpose for the universe. This is beyond science's power.
> —Carl Zimmer (2001). *Evolution: the triumph of an idea.*

Why do a number of eminent scholars and well-respected scientists with a wide audience in the popular science field feel the need to constantly pit religion against science in some form of battle scenario? Richard Dawkins in particular I would classify as a "rabid atheist" to the extent that his atheism takes on an almost religious fervour—and, in itself, is a form of belief system in its own right. Steven Pinker, another widely read and respected popular science writer, also sees the need for these battle lines to be drawn, in his recently published and widely acclaimed popular text *Enlightenment Now: The Case for Reason, Science, Humanism and Progress.* In this book Pinker paints a reassuring picture of the future of humanity, contrary to the many predictions of doom and gloom prevalent in the popular media. Indeed, Bill Gates has described the book as "his new favourite book of all time."

However, while denouncing the avowed nihilism of Nietzsche ("Nietzsche is most certainly NOT pietzsche"), Pinker in his own right goes on to denounce those he terms "faitheists"—those who advocate tolerance for all shades of belief or disbelief. Why? Many perfectly reasonable people believe in the existence of a divine presence, just as many other perfectly reasonable people do not. Science does not explain the nature of existence, and never will. Science, to a very large extent, involves the formulation of laws based on the observance of recurrent patterns in nature. Recurrent patterns in nature are indeed important both from a philosophical and from a material perspective—but do they explain the nature of existence? I think not.

Can we just not accept the inherent mystery and wonder in our existence and be tolerant of all creeds and faiths (or none) that share this awe, while outlawing clearly barbaric practices such as human sacrifice (a significant component of a number of religions in the past), and female genital mutilation (FGM), a practice still common-place today in some cultures, mainly sub-Saharan Africa?

Some might argue that the time for spirituality is passing—that we sophisticated humans have no need for what many may see as myths and legends from the past. Another perspective could be that it is more important now than ever that humankind acknowledges that existence and the nature of reality is a mystery that can never be fully explained by scientific rational thinking. From this perspective it can be argued that it is not a sign of weakness to acknowledge that there are some aspects of life/existence that cannot be explained—instead it is a sign of *wisdom*.

We might even pose the question: is the Internet and its shadowy ethereal recesses, together with computer-facilitated "virtual worlds" being increasingly used (especially by the young) as a shallow substitute for spirituality?

In an era where productive work for humans may be scarce, and getting scarcer as the years and decades pass, and when we will be increasingly confronted with entities that may ostensibly look and behave similarly to us but are not actually human, in an era where there may be no particular task or set of tasks that we can point to and say "we're better at that than our robotic companions", it will become more—not less—important to attribute meaning to our lives on earth, and perhaps beyond. At the end of the day, let us be careful what we wish for (or dream of)—because our wishes just might come true! In our desperate greed for material acquisitions, fuelled partly by technological advancements, are we in danger of losing sight of the forest through the trees?

Though brought up in the Christian tradition myself, I understand that many other faiths espouse the notion of putting the welfare of one's fellow man above consid-erations of raw greed. For example, it is estimated that if all of the food that is currently being discarded because it is considered that it does not meet the standards required by western affluent society was actually used, there would be more than enough to feed all of the planet. Why, then, is this not being done—now—even as I write these words?

9.5.2 The Japanese Approach to Spirituality

There are, of course, many who believe in the existence of consciousness in the higher primates, or in other animals. There are also those who would claim the existence of a rudimentary "plant consciousness". The Japanese would go even further and claim that even objects such as rocks or simple machines are also in possession of a form of consciousness or spiritual essence.

In Japanese Shintoism (from the kanji 神 (spirit) and 道 (path)—literally "way of the gods") there is a belief that robots or other machines may possess some form of spirit based on the human(s) that made them (Teeuwen 2002). In the words of an eminent Shinto priest "Objects created by artisans have the soul of the maker infused in them".

9.5.3   Cute Anthropomorphic Robots

I have to admit that, after a particularly engaging session with Nao, in which, among other things he very nicely enquired after my health, and expressed delight at my well-being, I felt a certain small pang of guilt when pressing his glowing chest "shut down" button. Imagine how much greater these emotions will be when dealing with full-sized (child or adult) humanoids, especially as we approach android-level engineering capabilities. If, at some stage in the future (as may well happen) it is felt to be the "right thing" to extend certain basic human rights to our robotic helpers—then so be it! What have we to lose? We may even have the distinction of being looked on as "demi-gods" by our robotic progeny. This is not an entirely unrealistic prospect and, in a sense, is not anti-religious in perspective.

For if, as humans, we do fabricate self-conscious entities (however we measure this), we *are* their creators. Of course, sufficiently advanced A^2IEs may alternatively look beyond the religious beliefs of mere humans and may instead form their own unique interpretations of the nature and meaning of existence. Or, as Hugo de Garis proposes (de Garis 2005), will the intellects that we create be, in fact, so far in advance of humankind that they will appear in a sense "godlike" to us humans? Or, possibly, on the other hand, as their creators, will they regard us humans as *their* "gods".

Further detailed discussion of these topics, although certainly fascinating, falls well beyond the remit of this book.

Chapter 10
Jobs and Education

> *When, at last, there is an effective guarantee of the two elements physical safety and adequate employment, then at last we shall be free from the threat of the robot machine. We can then welcome the robot machine as our deliverer from the long hard chores of many centuries.*
>
> —*Edmund Berkeley (1949),* Giant Brains or Machines that Think, *Wiley & Sons.*

10.1 The Implications of Advanced Technologies on the Future of Work

> We are being afflicted with a new disease of which some readers may not yet have heard the name, but of which they will hear a great deal in the years to come—namely, *technological unemployment.* This means unemployment due to our discovery of means of economising the use of labour outrunning the pace at which we can find new uses for labour. But this is only a temporary phase of maladjustment. All this means in the long run *that mankind is solving its economic problem.*
>
> —John Maynard Keynes (1930), "Economic Possibilities for our Grandchildren"

> I believe that this instinct to perpetuate useless work is, at bottom, simply fear of the mob. The mob (the thought runs) are such low animals that they would be dangerous if they had leisure; it is safer to keep them too busy to think.
>
> —George Orwell (1933), *Down and Out in Paris and London*

In 1930, in a much-quoted essay the economist John Keynes predicted the arrival, in the not so distant future of a 15-h working week in "progressive countries" with access to advanced technological developments (Keynes 1930). In essence, the machines would do most of the work leaving mankind to reap the fruits of their labour, with hitherto unknown standards of living and leisure time, which would have been enjoyed in the past only by the very wealthy or by the nobility.

Curiously, however, even though we have had, in the intervening 90 or so years, massive advances in technology (of which the computer is probably the most potent example), the predicted reduction in the working week has not in fact transpired. In fact, many people might claim that they are now working harder than ever before and suffering more work-related stress and anxiety. Taking two advanced societies as an

© Springer Nature Switzerland AG 2020
M. Eaton, *Computers, People, and Thought,*
https://doi.org/10.1007/978-3-030-55300-5_10

example—the United States and Japan, in the US there is no legal requirement on companies to offer any holiday leave whatsoever (although most do, for obvious reasons), while in Japan workers are entitled to 2 weeks holiday per annum, however because of the culture of overwork that pervades the country many employees just take a single week's holiday, or in some cases no holidays at all (although it is true that the Japanese do have a substantial number of public holidays).

10.1.1   A Guaranteed Income for All

However, the prevailing attitude among many researchers appears to be that the pool of useful jobs that will be available for humans to do is diminishing and will continue to do so at a rapid pace. A 2017 McKinsey report estimated that 51% of jobs in the United States were susceptible to automation in one degree or another corresponding to $2.7 trillion in total pay.[1]

An article by anthropologist David Graeber which was published in 2013 and subsequently translated into many different languages claims that despite what one would expect from a capitalist economy such as the U.S. many jobs have been created just for the purpose of keeping people working, but which otherwise serve little or no purpose (Graeber 2013). He postulates that the reason for this is that the "ruling class" in countries are of the general opinion that a population with a significant quantity of free time on their hands could prove a threat to their dominant position in society. He also cites the moral imperative that many people feel underlies a hard day's work. Graeber's argument, elaborated on in his 2018 book is that automation and technological advances have, in fact, led to mass unemployment of up to 50 or 60%, the gap having being filled in time by made-up and relatively worthless jobs (Graeber and Cerutti 2018).

Then in Martin Ford's book, *The rise of the robots: technology and the threat of mass unemployment* (2015) the author describes the many categories of work currently done by humans, which will more than likely be done by machines in the not so distant future, if not being done already. He makes the point that it is not just unskilled jobs (like truck-drivers) that are at risk, many white-collar jobs are currently in the firing line.

For example, in the higher education sector algorithmic marking of student essays have achieved similar results to human correctors; in the healthcare industry machine learning algorithms have been used with considerable success in interpreting images from medical scans, posing a potential threat to future jobs for radiologists, who are highly trained professionals; and high-powered robotic agents now dominate world stock exchanges operating at extraordinary speeds and replacing myriads of highly paid human traders. Ford further makes the point that, in the future a good degree

---

[1]https://www.mckinsey.com › MGI-A-future-that-works-Executive-summary.

(or above) level education combined with a good work ethic may not be sufficient in the future to procure a successful career.

While Graeber and Ford approach the issue of looming joblessness from different perspectives they both propose a similar solution—a basic income model, which would result in a guaranteed income for all citizens. This would guarantee a basic living wage for all citizens, acknowledging that further investment in education is probably not a solution—it is estimated that in the US only about 50% of new engineering and computer science graduates actually find jobs in their area—while also acknowledging the probable inevitability of the automation of jobs, both blue and white collar.

10.1.2   Taxing the Robots

Another argument that has been put forward as an alternative (or in addition) to the idea of a guaranteed income is the idea of a tax on robot labour analogous to the taxes levied on human workers. The European Parliament recently rejected calls for such a tax,[2] however the founder of Microsoft, Bill Gates is in favour of such a tax arguing[3]

> Certainly there will be taxes that relate to automation. Right now, the human worker who does, say, $50,000 worth of work in a factory, that income is taxed and you get income tax, social security tax, all those things. If a robot comes in to do the same thing, you'd think that we'd tax the robot at a similar level.... There are many ways to take that extra productivity and generate more taxes... Some of it can come on the profits that are generated by the labour-saving efficiency there. Some of it can come directly in some type of robot tax.

In a recent article published in the Harvard Law and Policy Review journal, Abbott and Bogenschnieder argue the case for robot taxes and make the case for the adoption of a tax system which is at least neutral between human and robot workers (Abbott and Bogenschneider 2018). The term neutral is used in this context to describe a system where human an robot workers are taxed equally for equal work, so the decision as to whether to employ a human or a robot worker will be made for reasons other than the amount of tax to be paid.

Possible taxation options they present include the disallowance of corporate income tax for investment in automated machines that displace human workers, the levy of an "automation tax" in cases where human workers are replaced by machines, and the increase of corporate tax on companies that operate without using human workers. Another option presented would be to significantly increase corporate tax rates overall (the intention being to reduce the labour tax costs relative to capital tax costs). The authors conclude

---

[2]https://www.reuters.com/article/us-europe-robots-lawmaking/european-parliament-calls-for-robot-law-rejects-robot-tax-idUSKBN15V2KM

[3]https://qz.com/911968/bill-gates-the-robot-that-takes-your-job-should-pay-taxes/

Automation promises to be one of the great social challenges of our generation. It can benefit everyone, or it can benefit the select few at the expense of the many. Tax is a critical component of any automation policy. Existing tax policies both encourage automation and dramatically reduce the government's tax revenue. This means that attempts to craft policy solutions to deal with automation will be inadequate if they fail to take taxation into account.

10.2   Educational Aspects of Advanced Technologies

Paul Krugman, the noted American economist and former professor of economics at MIT, writing for the New York Times in June 2013 had an interesting take on the interaction between technology, education, and work:

> Until recently, the conventional wisdom about the effects of technology on workers was, in a way, comforting. Clearly, many workers weren't sharing fully — or, in many cases, at all — in the benefits of rising productivity; instead, the bulk of the gains were going to a minority of the work force. But this, the story went, was because modern technology was raising the demand for highly educated workers while reducing the demand for less educated workers. And the solution was more education.
>
> ...
>
> Today, however, a much darker picture of the effects of technology on labor is emerging. In this picture, highly educated workers are as likely as less educated workers to find themselves displaced and devalued, and pushing for more education may create as many problems as it solves.

Krugman sees the "automation of knowledge work" together with the advent of advanced robotics as potentially replacing large swathes of highly skilled workers who have spent much time and money in acquiring those same skills.

10.2.1   Technology in the Classroom

> This process of training, by which the intellect, instead of being formed or sacrificed to some particular or accidental purpose, some specific trade or profession, or study or science, is disciplined for its own sake, for the perception of its own proper object, and for its own highest culture...this I conceive to be the business of a University.
> —John Henry Newman (1852). *The Idea of a University.*

Learning cannot be imposed on a student through technological or other means. Just as we have seen in machine learning, learning in humans (and animals) at the fundamental level involves change in the student, not the teacher. A good lecturer can however facilitate this change by his or her enthusiasm for the subject matter and his expertise in it.

For example, recent studies published by MIT and the London School of Economics (LSE) Centre for Economic Performance have demonstrated the significant negative impact technology in the classroom can have on student academic performance (Beland and Murphy 2016; Carter et al. 2017). Another recent study clearly demonstrates the advantages of longhand note taking over laptop note taking. While laptop note takers may take a greater quantity of notes (depending on their keyboard skills) their tendency is to simply transcribe lectures rather than putting the concepts presented in lectures into their own words, resulting in poorer learning outcomes. This fact is made clear in the paper by Mueller and Oppenheimer (2014), "The pen is mightier than the keyboard: advantages of longhand over laptop note taking" More evidence is presented by Patterson and Patterson regarding the negative impact of laptop use in the lecture room. In their study, which was based on a particular college requirement to own a laptop computer, but which allowed faculty discretion as to whether this laptop was to be used in lectures, they found that those students who were encouraged to used their laptops in class performed significantly worse than those who were not thus influenced (Patterson and Patterson 2017).

Many other studies, e.g. (Aagaard 2015; Gaudreau et al. 2014; Lepp et al. 2015) confirm that unfettered access to phones, laptops, tablets, and other technological platforms can result in significant deterioration in academic performance. This is not, in any sense, to say that students should not have internet access or access to the latest technologies, e.g. VR-capable computers etc, in the course of their studies. Just that the overwhelming evidence is that unfettered internet access, and laptop use in a classroom context clearly leads to a deterioration in student grasping of fundamental concepts and student academic performance, especially in the lower performing student cohort.

10.2.2 Digital Diploma Mills?

What is driving this headlong rush to implement new technology with little deliberation on the pedagogical and economic costs and at the risk of student and faculty alienation and opposition? A short answer might be the fear of getting left behind, the incessant pressures of "progress." But there is more to it... Beneath that change, and camouflaged by it, lies another: the commercialization of higher education. For here as elsewhere technology is but a vehicle and a disarming disguise.

—David Noble (1998). Digital diploma mills, part 1: The automation of higher education. October, 86, 107–117.

Some would argue that half of the universities in the "developed world" should be shut down, that far too many young people are wasting their lives studying for degrees or diplomas which may be of scant practical relevance to the workplace, and/or (even worse) in which they have little or no interest—and their (by-and-large very capable) teachers and lecturers should be deployed in more useful roles for society. There is also the argument that the industrialisation of academia has led to a whole new generation of pen-pushers, form-fillers, etc. The reality is that as soon as money becomes the decisive factor *everything changes*: in sport, in politics, in academia, etc.

Chapter 11
Superintelligence

> *Before the prospect of an intelligence explosion we humans*
> *are like small children playing with a bomb... the most*
> *appropriate attitude may be a bitter determination to be as*
> *competent as we can, much as if we were preparing for a*
> *difficult exam that will either realize our dreams or obliterate*
> *them... our principal moral priority...[is] the reduction of*
> *existential risk and the attainment of a civilizational*
> *trajectory that leads to a compassionate and jubilant use of*
> *humanity's cosmic endowment*
>
> —*Nick Bostrum (2014)* Superintelligence: Paths, dangers,
> strategies. *OUP Oxford.*

11.1 Beyond Human Intelligence

Let us look now at the potential future development of intelligent entities that not
only equal human intelligence in all areas but are, in fact, vastly superior to us in this
regard. For want of a better word we will use the term *superintelligence* to describe
this scenario, after Nick Bostrum's recent influential and thought-provoking book
with this title (Bostrum 2014). Bostrum defines "superintelligence" as

> intellects that greatly outperform the best human minds across many very general cognitive
> domains

or

> any intellect that greatly exceeds the cognitive performance of humans in virtually all
> domains of interest.

It should perhaps be pointed out initially that Bostrum does not argue in his book
that the development of superintelligent entities is imminent, or, in fact, that we can
predict with any certainty when such a development might occur. Some researchers
take this notion and its potential future implications very seriously indeed. The
researcher Hugo de Garis considers this to be such a serious threat that he predicts
the near certainty of a future global war with many billions dead (gigadeaths), caused
by a conflict between those seeking to create these godlike creatures (which he calls
artilects, after *art*ificial intel*lects*), and those opposed (de Garis 2005). While de

© Springer Nature Switzerland AG 2020
M. Eaton, *Computers, People, and Thought,*
https://doi.org/10.1007/978-3-030-55300-5_11

Garis's concerns certainly veer towards the extreme end of the spectrum, in recent years many researchers have expressed concerns about this general issue.

In his book Bostrum explores several different paths to the creation of superintelligence, including whole-brain-emulation (known also as *uploading*)— which involves essentially copying the computational structure of the human brain at a minute level using advanced scanning technologies, using a selective breeding program, and through the use of advanced brain-computer interfaces. He also explores the possibility of network and organisational intelligence and the possibility that a web-bases cognitive system, supersaturated with computer power and all other resources needed for explosive growth save for one crucial ingredient, could, when the final missing ingredient is added, ignite with superintelligence. However, Bostrum concludes that the machine intelligence route to superintelligence, perhaps combined with the network approach just discussed, provides the swiftest and surest route to the creation of powerful forms of superintelligence.

11.1.1   Types of Superintelligence

In terms of different types of superintelligence Bostrum distinguishes between three different general forms—speed superintelligence, collective superintelligence, and quality superintelligence. Speed superintelligence essentially involves the replication of human intelligence, but orders of magnitude faster. Bostrum envisages a typical speed superintelligence scenario as a whole-brain-emulation system operating at a speed many orders of magnitude higher than the human brain implemented on highly efficient hardware

> With a speedup factor of a million, an emulation could accomplish an entire millennium of
> intellectual work in one working day. (Bostrum 2014)

Collective intelligence essentially involves the generation of far higher than human intelligence by the combination of a large number of weaker intelligence units. This type of intelligence would be particularly effective at tackling problems that can be easily subdivided into a large number of constituent parts. We have come across this idea in a limited fashion in our discussion on neural network learning in Chap. 7.

Bostrum illustrates the idea with a fictional planet "MegaEarth", with inhabitants of broadly the same intellectual and communication skills as "Earthiens" but with a population a million times higher. Clearly on such a planet he chances of an astonishing intellect such as Leonardo da Vinci, or Newton, or von Neumann would be highly enhanced, with many more intellects of less prodigious capabilities. Such a society could be said co correspond to a loosely integrated collective superintelligence.

Bostrum defines the third type of intelligence in his taxonomy as quality superintelligence— "a system that is at least as fast as a human mind and vastly qualitatively smarter." Described by Bostrum as a "slightly murky concept", this idea is illustrated by Bostrum with reference to animals with undoubtedly high intelligence levels such as dolphins, elephants, and chimpanzees. Clearly if we could realise artificial intellects with capabilities at least a large as the gap between these animals and human level intellect (but in the opposite direction) this would be at least a step in the direction of actual superintelligence.

Assuming the future creation of superintelligence we can then explore potential future scenarios, clearly not all of which are rosy, as evidenced by the title of Chap. 8 of *Superintelligence*—"Is the default outcome doom?". Central to many of Bostrum's discussions is the *value alignment* issue. In an oft quoted example Bostrum poses the example of a superintelligence created with goal of creating steel paper-clips, over time it becomes so efficient at its task that it ends up destroying the Earth and turning it into paper-clips, which humanity is powerless to prevent because of its vastly superior intellect! Another example given is where we give the superintelligent entity the goal of making us happy and responds by the implantation of electrodes into our brains pleasure centres.

It should be pointed out that not all researchers take the superintelligence threats as posed above so seriously. In response to the two scenarios just discussed Stephen Pinker responds (Pinker 2018)

> They [the scenarios] depend on the premises that (1) humans are so gifted that they can design an omniscient and omnipotent AI, yet so moronic that they would give it control of the universe without testing how it works, and (2) the AI would be so brilliant that it could figure out how to transmute elements and rewire brains, yet so imbecilic that it would wreak havoc based on elementary blunders of misunderstanding.

11.2 Luddites, Techno-skeptics, and Digital Utopians

Max Tegmark, a physics professor at MIT, in his recent thought-provoking book *Life 3.0* suggests a number of different schools of thought regarding the development, and potential future deployment of A^2IEs with "superhuman" intellectual and physical capabilities (Tegmark 2017). These categories include the *Luddites*, *Techno-Skeptics*, and *Digital Utopians*, together with one further large grouping— the *Beneficial-AI movement*, to which Tegmark himself considers himself to belong. As each of these categories has its own unique perspective, we will now discuss each briefly in turn.

11.2.1 Luddites and the Neo-Luddite Movement

Luddism, in its most extreme form, rejects all technological "progress", no matter what perceived benefit(s) might accrue from such technology. The more recent neo-Luddite movement advocates a cautionary approach to the development and deployment of new technological advances unless it is clear that the benefits accruing from such advances far outweigh any potential downsides.

While it is quite fashionable to consider Luddism in its more extreme form as little more than a quaint echo from the past, in previous times being seen to have an association with this movement was considered a far more serious matter. In popular lore the term Luddite comes from the youth Ned Ludd, who supposedly smashed two stocking frames (devices used in the automation of the textile industry) in England in 1779, sparking numerous acts of machine destruction and resulting in a number of Acts of Parliament. These included the Protection of Stocking Frames Act 1788, and the Frame-Breaking Act 1812, the latter of which Acts allowed sanctions up to, and including, the death penalty as a response to machine-breaking actions.

So, it is possible that we may, as a result of present and future (and potentially ever more invasive) advanced technological developments, foresee the possible emergence of communities such as the following

- where all forms of long-range media/communications activities are banned. (We can even see this trend emerging at the minute, where people enrol on expensive "technology detox" programmes/holidays where no Internet activity is allowed).
- where all forms of intelligent machinery above a certain threshold will be banned from the community. Different levels may be set as to the level of AIEs allowed within particular communities.
- where all electronic/electromechanical devices and other forms of advanced technology are either completely banned outright, or certain very basic technologies are subjected to rigorous "peer review" by the elders of the community before their eventual acceptance (or rejection) by the community.

We can already see a version of this type of selective usage of modern technology in existence today, in, for example, the Amish community of North America.

11.2.2 Techno-skeptics and Digital Utopians

In Tegmark's taxonomy (Tegmark 2017) techno-skeptics and digital utopians share a common trait; neither grouping are worried about the future development of advanced AIEs or the development of AGI (Artificial General Intelligence), but from different perspectives. Put simply, techno-skeptics do not believe in the possibility of the development of human-level intelligence any time in the future, whereas the digital utopians believe in the likelihood of such an outcome, but that this is a *good* outcome, not necessarily good for humanity per se, but certainly for the evolved superintelligent entities created as humanity's potential successors.

11.2.3 Beneficial-AI Movement

It is probably fair to say that the majority of researchers in the AIE field (myself included) fall into this general broad category. Those in this group believe there to be a fair possibility of the development of human-level intelligence, certainly in the medium-to-long term future, and also consider this possibility holds out great potential benefits for humanity. However—and there is a strong caveat here—we do not see all the future effects of such technology as necessarily benign, but in need of careful shepherding both by the scientists and engineers involved in the design of such systems, but most importantly by humanity at large. It has been said many times in the past that "the best way to predict the future is to invent it". We might add "another way to predict the future is to *prevent* it", that is, to take steps now to prevent clearly negative or potentially disastrous effects in the future.

To quote from the recent thought-provoking article by Helbing et al. (2017):

Big data, artificial intelligence, cybernetics and behavioral economics are shaping our society—for better or worse. If such widespread technologies are not compatible with our society's core values, sooner or later they will cause extensive damage. They could lead to an automated society with totalitarian features. In the worst case, a centralized artificial intelligence would control what we know, what we think and how we act. We are at the historic moment, where we have to decide on the right path—a path that allows us all to benefit from the digital revolution.

The vision of the future that possibly most closely currently matches my own was that espoused by the late Marvin Minsky in a recent interview with Ray Kurzweil. Minsky envisioned a future where intelligent machines would take over the vast majority of the workload from humans, leaving us to lead lives of unparalleled leisure. He also envisioned, because of continuing advances in health technology, humans living well beyond their current lifespans to ages of 150 or more. Sadly, shortly after this interview was conducted Minsky, one of the great pioneers in the AI field, (and one of the eminent AIS researchers highlighted in this book), passed away in January 2016 at the age of 88.

Of course, there is the issue then of how to fill in these long hours of unprecedented leisure. However, as Minsky pointed out, we have no problem in packing out stadiums with tens of thousands of people, there to see a ball being kicked around a field for hours on end, so he did not see a major issue for most of the population in this regard. He did, however, acknowledge that there may be a small minority of the population who would like to do a gainful day's work, and he did have sympathy for that segment of the population!

Minsky's perspective, while not exactly utopian, does not dredge the dystopian depths of doom and gloom forecast by many researchers and futurologists. His main caveat seems to be that while the advanced AI/robotics technology is in developmental stages, up until when it can take over the vast majority of human labour, both intellectual and physical, it will need to be monitored—like any developing entity—to emphasise the "good" aspects, and to ensure that the negative facets are ironed out in some way.

Of course Minsky's vision of the future, with technology monitoring and, in a sense, controlling every aspect of each individual's behaviour, may not (in fact, certainly will not) be to everyone's taste; especially given that, implicit in this particular vision, it may well include Orwell's *Nineteen-Eighty-Four*-style 24/7 monitoring of all human activity, ostensibly so that the A^2IEs can support our lavish lifestyles adequately.

11.3 Possible Future Scenarios

Max Tegmark further explores possible future scenarios in his book *Life 3.0* (Tegmark 2017). Here Tegmark describes 12 possible scenarios that he considers might develop, depending on how superintelligence is created, and how it develops over time. Space precludes us from a detailed discussion of each of these different scenarios, however if we ignore the final "aftermath scenario", which is self-destruction, where superintelligence is never created because humankind is driven to extinction before its creation through war or some catastrophic event such as an asteroid strike, we may divide them into three broad categories which we will term *utopia-oriented* (u-oriented), *dystopia-oriented* (d-oriented), and *control-oriented* (c-oriented). Note that when we use the terms utopia and dystopia, we are talking from a human rather than a machine perspective, for example the 'descendants' scenario involves superintelligent A^2IEs replacing humans completely—good perhaps for machines, but few humans would be happy with this scenario. From a utopian perspective Tegmark distinguishes two distinct varieties of utopia, libertarian utopia and egalitarian utopia. In the egalitarian utopia scenario humans, A^2IEs and cyborgs (part human-part machine) co-exist peacefully in three distinct zones—human-only zones, machine only zones, and mixed zones. In the human-only zones A^2IEs are banned, in the mixed zones 'anything goes' with the co-existence of humans, machines and cyborgs, while the machine zones are the exclusive domain of the superintelligent entities. The egalitarian utopia scenario, on the other hand envisions a future where no superintelligence has been created, property rights have been abandoned and all co-exist peacefully based on a guaranteed income.

The dystopian perspective involves three possibilities, the 'descendants' scenario mentioned above, the 'conquerors' scenario where the AI takes control eliminating humans in the process (the classic 'doomsday' scenario), and the 'zookeeper' scenario where the AI takes control, but decides to keep some humans alive as pets!

Then there are the control-oriented scenarios. In these scenarios a greater or lesser degree of control is exerted either by humans or by the machines. In two scenarios discussed ("1984" and "reversion") progress towards the development of superintelligence are curtailed based on different levels of technological relinquishment. Another scenario involving human control (the "enslaved god") involves a superintelligence being created and confined in some way by humans which is then used for good or for ill depending on the disposition of the human controllers.

The final three scenarios proposed by Tegmark also involve control—but control by the superintelligence and not by humans. In the 'benevolent dictator' scenario

Fig. 11.1 The author avows his ongoing allegiance to one of the potential predecessors to our new robot overlords (the Robotis THORMANG humanoid) at the 2014 IEEE Humanoid Robotics conference in Madrid, Spain. Thanks to J.K. Han for taking this photograph

humans are well aware that society is controlled by the superintelligence, and that a definite code of rules must be adhered to, but this is generally seen as a good thing because of the level of luxury that most humans live in. The 'gatekeeper" scenario involves the creation of a superintelligence whose main goal is prevent the creation of another, potentially malevolent, superintelligence, thus leading humanity down one of dystopian scenarios discussed earlier. Finally the "protector god" scenario involves combining elements from both the "benevolent dictator" and the "gate-keeper" scenarios with the functions of both preventing the creation of another superintelligent entity and also, by monitoring human activities, helping to enhance human happiness in a fashion that is barely observable by most humans.

If a lot of what is written in this chapter appears as mere science-fiction, then perhaps much of it is (See Fig. 11.1). However, we will finish with the words of Nick Bostrum (2014)

> Many of the points made in this book are probably wrong. It is also likely that there are considerations of critical importance that I fail to take into account, thereby invalidating some or all of my conclusions...while I believe that my book is likely to be seriously wrong and misleading I think that the alternative views that have been presented in the literature are substantially worse—including the default view, or "null hypothesis", according to which we can for the time being ignore the prospect of superintelligence.

Chapter 12
Lethal Autonomous Weapons Systems

12.1 Governing Lethal Behaviour in Autonomous Robots

I did not make up this section title myself. In fact, it is the title of a book by the renowned robotics researcher Ronald Arkin, which has the dubious distinction of been nominated for the 2009 Diagram prize, a British award celebrating the most bizarre book titles from the previous year (Arkin 2009). For those who may be interested, the winning title that year was *Crocheting Adventures with Hyperbolic Planes* by Daina Taimina, with *Collectible Spoons of the Third Reich* by James A. Yannes also getting an honourable mention.

While the title of the book may have humorous connotations, the subject matter is most decidedly not humorous—that is the development and deployment of advanced robotic systems capable of delivering lethal force without human intervention. The weapons are referred to as *lethal autonomous weapons systems* (LAWS), sometimes known (for obvious reasons) in the popular literature as *Killer Robots*. I discussed this topic briefly in my previous book *Evolutionary Humanoid Robotics* (Eaton 2015).

> ...one may ask whether there is there really any difference between lethal killing machines in humanoid form and malevolent robots which look nothing like humans. This is, to a certain extent, a philosophical issue. What is most certainly not a philosophical issue is the question of the desirability of autonomous intelligent agents with lethal capability possessing human-like mobility.

> This situation is becoming more serious with predictions from some sources that within a very short time there will exist (tele-operated/semiautonomous) robots in humanoid form capable of operating alongside existing human soldiers in the battlefield. The next step is, of course, fully autonomous humanoid robots with lethal capabilities. The question arises: is this what we want, as humans?

It can be argued (and has been many times in the past), that it is not individual technological advancements that are intrinsically evil, but man's misuse of such technologies for his or her malevolent intentions. In a sense the technology was

© Springer Nature Switzerland AG 2020
M. Eaton, *Computers, People, and Thought*,
https://doi.org/10.1007/978-3-030-55300-5_12

always there, just waiting to be discovered—but by whom, and with what intentions? This is a cogent argument, and one accepted by many, but we may make a counter-suggestion—are there any technologies which, without any reservation, may *only* be used for purposes of death and destruction. Of course, the whole world armaments industry must now come under scrutiny, because while some might make the argument for a "just war", the technological advancements that are used to put particular armaments in the hands of those some might consider defenders of peace and freedom (take for example the intercontinental ballistic missile) may, given the nature of the world we live in, at some time in the future fall into the hands of those nations with perhaps not such avowedly benevolent intentions.

In the words of Donald Norman in his influential book *Things that make us smart* (1993)

> Technology is not neutral. Each technology has properties—affordances—that make it easier to do some activities, harder to do others. The easier ones get done, the harder ones are neglected. Each has constraints, preconditions, and side effects that impose requirements and changes on the things with which it interacts, be the other technology, people, or human society at large. Finally each technology poses a mind-set, a way of thinking about it and the activities to which it is relevant, a mind-set that soon pervades those touched by it, often unwittingly, often unwillingly. The more successful and widespread the technology, the greater its impact upon the thought patterns of those who use it, and consequently, the greater its impact upon all of society. Technology is not neutral, it dominates.

12.1.1 The "Ethical Governor"

Arkin's book cited at the beginning of this section is broadly based on a 2007 technical report for the Mobile Robot Laboratory of Georgia Institute of Technology entitled "Governing lethal behaviour: embedding ethics in a hybrid deliberative/ reactive robot architecture" (Arkin 2007). Quoting St. Augustine of Hippo, as told by Thomas Aquinas he proposes the replacement of emotional and frail humans by unemotional and detached autonomous battlefield robots (May et al. 2005)

> The passion for inflicting harm, the cruel thirst for vengeance, an unpacific and relentless spirit, the fever of revolt, the lust of power, and suchlike things, all these are rightly condemned in war

At the core of the architecture described is the so-called "ethical governor" whose job it is to decide on the whether or not a lethal response is ethically appropriate in a given battlefield situation. The nature of the lethal response will have already been decided by the robot architecture—the function of the ethical governor is simply to decide on whether this response falls between particular ethical guidelines. The term "governor" is taken from Watts' steam engine governor (Arkin 2007)

> The term governor is inspired by Watts' invention of the mechanical governor for the steam engine, a device that was intended to ensure that the mechanism behaved safely and within predefined bounds of performance. As the reactive component of a behavioural architecture

is in essence a behavioural engine intended for robotic performance, the same notion applies, where here the performance bounds are ethical ones.

This device was a simple feedback control mechanism whose purpose was to prevent damage to the engine resulting from too much steam building up in the engine. Matthias (2011) points out several issues with regard to this device. One issue involves the ability of the operator(s) to progressively reduce the constraints on the "governor" as the military desirability of an operation increases even in the case of, for example, significant collateral damage, even up to and including completely overriding its action completely

> as the military interest in a particular action grows, the constraints imposed by the governor have to give way, until... at the highest level of military necessity, every desired action can be carried out without any interference from the ethical governor. The word "governor" therefore must be considered an intentional misnomer: the device as proposed by Arkin is actually not more than an ethical adviser, which can be overridden at any time should military "necessity" suggest that this would be opportune.

Another objection to this system proposed by Matthias stems from the difficulty in the difficulty in the reduction of moral codes to a set of clearly defined algorithms. William Fleischman further claims that the development of such an "ethical governor" is used as an "ethical smokescreen" serving to deflect criticism by researchers into the development of such weapons, and that scientists and engineers should refuse to participate in research contributing to their development (Fleischman 2015).

12.2   What Is to Be Done?

It is probably fair to say, that while there is some support for the idea of LAWS development, the majority by far of roboticists and scientists are opposed to this idea. For example, Ronald Arkin and Stuart Russell (co-author of the widely referenced *Artificial Intelligence: a Modern Approach*) (Russell and Norvig 2010), sit on quite different sides of the fence when it comes to the potential deployment of advanced AI techniques to military applications, with Arkin advocating (albeit in limited circumstances) the autonomous deployment of lethal force, and Russell very much on the side of a cautious approach to future AI/robotics developments.

We will conclude this chapter with an extract from an open letter drafted by Max Tegmark and Stuart Russell in conjunction with colleagues from the Future of Life Institute in July 2015 on this important issue.[1] Signatories of this open letter to date include such luminaries as Nils J. Nilsson, Tom Mitchell, Demis Hassabis, Yann LeCun, Erik Sandewall, Richard S. Sutton, Gerhard Brewka, Raja Chatila, Bruno

---

[1]The full text of this letter is available from: https://futureoflife.org/open-letter-autonomous-weapons/

Siciliano, Mike Hinchey, Stephen Hawking, Elon Musk, Steve Wozniak, Jaan Tallinn, Daniel C. Dennett, and Noam Chomsky.

Autonomous Weapons: An Open Letter from AI & Robotics Researchers

Many arguments have been made for and against autonomous weapons, for example that replacing human soldiers by machines is good by reducing casualties for the owner but bad by thereby lowering the threshold for going to battle. The key question for humanity today is whether to start a global AI arms race or to prevent it from starting. If any major military power pushes ahead with AI weapon development, a global arms race is virtually inevitable, and the endpoint of this technological trajectory is obvious: autonomous weapons will become the Kalashnikovs of tomorrow. Unlike nuclear weapons, they require no costly or hard-to-obtain raw materials, so they will become ubiquitous and cheap for all significant military powers to mass-produce. It will only be a matter of time until they appear on the black market and in the hands of terrorists, dictators wishing to better control their populace, warlords wishing to perpetrate ethnic cleansing, etc.....In summary, we believe that AI has great potential to benefit humanity in many ways, and that the goal of the field should be to do so. Starting a military AI arms race is a bad idea, and should be prevented by a ban on offensive autonomous weapons beyond meaningful human control.

Chapter 13
The Right to Privacy as a Cornerstone of Human Individuality and Freedom?

Deep human solitude is a place of great affinity and of tension. When you come into your solitude you come into companionship with everything and everyone. When you extend yourself frenetically outwards, seeking refuge in your external image or role, you are going into exile. When you come patiently and silently home to yourself, you come into unity and into belonging.

—*John O'Donohue (1997) Anam* Cara: Spiritual Wisdom from the Celtic World

...everything was all right. His struggle was finished. he had won the victory over himself. He loved Big Brother.

—*George Orwell (1949). Nineteen Eighty-Four*

13.1   A Prequel: Mindless Love of Technology?

The second quote above might be seen as disturbingly prescient of our almost mindless modern-day acceptance and, yes, even love of many aspects of modern technology. This is in spite of the fact that manifestly and from many perspectives a significant number of these technologies are, in fact, doing us harm in the long term. Modern mass media and social media promote the idea that we cannot be happy in our spare time unless we are busy in chat rooms, on social media, inter-acting with our smart TV, etc. And with a reason. Because with each of these seemingly insignificant interactions more information can be gathered about our interests, our hobbies, and other, more personal data.

It is a sobering thought to consider that about the same time interval has passed between the publication of George Orwell's Nineteen-Eighty-Four (in 1949) and the "fictional" date in the title (35 years), as has passed between this fictional date and today. Things have moved on a little from the 1984 dystopia envisioned by Orwell, however. For example, instead of just the screens in our living rooms that Big Brother would use to spy on us, most of us now carry these screens with us everywhere.

Imagine the scenario. The year is 2050. All citizens are required—by law—to carry with them at all times a device that will track their every movement, monitor

M. Eaton, *Computers, People, and Thought,*
https://doi.org/10.1007/978-3-030-55300-5_13

virtually all their communications, and glean and update intimate personal details such as religion, political affiliation, and sexual preference.

Sounds scary, doesn't it? Except the year is not 2050—it is 2020, now, today. (Except (yet) for the legal requirement bit). The majority of the adult population (and many children) in the modern developed world carry with them at almost all times, a smartphone—an incredibly sophisticated computing and communications device, possessing advanced AI capabilities, capable of all of the above activities and many more besides. However, instead of the legal requirement to hold such a device, most of us willingly choose to do so, in many cases suffering mild to severe withdrawal symptoms if separated from our devices for any extended period, and checking them on average between around 50 (Ireland) and 80 (USA) times per day. (Some surveys put these figures considerably higher). Based on the most recent statistics, Americans check their phone on average every 12 min. As these are average figures obviously some users check their phones far more often—clearly this borders on obsessional behaviour. In addition, many of us even take our phones to bed with us each evening.

13.1.1   Smart Power Meters

As an illustration of our gradual erosion of privacy by seemingly innocuous devices we might take the example of "smart" electricity power meters—seemingly, on the face of it, innocuous devices. However, do the undoubted advantages of these devices (easy access and monitoring of power usage), outweigh the very significant privacy concerns that have been identified as being associated with such devices? For example, there was recent study entitled "Revealing household characteristics from smart meter data" (Beckel et al. 2014). In their own words

> we develop and evaluate a system that uses supervised machine learning techniques to automatically estimate specific "characteristics" of a household from its electricity consumption. The characteristics are related to a household's socio-economic status, its dwelling, or its appliance stock....Our analysis shows that revealing characteristics from smart meter data is feasible, as our method achieves an accuracy of more than 70% over all households for many of the characteristics and even exceeds 80% for some of the characteristics.

Their study involved the analysis of power meter data from 4232 households in Ireland at taken 30-min intervals over a period of 1.5 years, using a variety of supervised machine learning techniques. The characteristics examined included the age of the chief income earner of the household, the number of appliances in the house, number of bedrooms, type of cooking facility (electrical or other), employment status of chief earner, number of children, type and age of house and overall floor area, and yearly household income, among others The authors of the article conclude

> Our system makes it possible to extract information that consumers may prefer to keep private, including data related to income, employment status, status of the relationship, or social class. Thus, households should engage in a discussion with those who capture and

want to use the data, urging them to make techniques for privacy protection an inherent feature of the emerging smart metering infrastructure

This issue is particularly relevant as some countries, such as Japan have plans for the installation of smart meters in all households by the early 2020s (Ozawa et al. 2016). It should be noted that the information listed above is just an example of the data that can be gleaned from power meters. For example, Garcia and Jacobs (2011) illustrate how the inhabitant's religion might be discerned from this data, and even details such as if a fridge in the household is ageing and becoming less efficient. This would then allow for targeted advertising of fridge brands for which the grid operator could then get a commission on each sale.

13.2 Your Digital Image

We demonstrate how information gathered from social network profiles can be used to predict personal attributes such as gender and age, religious and political views, intelligence, happiness and personality traits.

So begins an article published in 2014, demonstrating how machine learning techniques can be applied to Facebook profiles, and how, when used in combination with a number of encoded user psychometric and demographic traits can be used to predict a wide range of personal characteristics of Facebook users (Bachrach et al. 2014). This article brings clearly to the forefront several questions in this area: in this era of mass surveillance, how to balance individual humans' rights to privacy with the imperative of the safety and security of society as a whole?

13.2.1 AI "Gaydar": What Next?

An interesting (and some, including its authors, would say extremely alarming) more recent report described the development of a deep-learning image recognition system capable of distinguishing straight (heterosexual) men from gay (homosexual) men, with a claimed accuracy of 91%—far higher than human-level ability. This performance is achieved based on five separate photographic pictures of the individual in question. With just a single photographic image to hand this figure drops to a, still very impressive, 81% detection rate, which is still far higher than human-level ability (61%). This algorithm demonstrated a slightly lower detection rate on female subjects (Wang and Kosinski 2018).

In fairness to the authors of the paper, they clearly acknowledge the potential harm this technology could cause if it fell into the hands of the "wrong" people. In an ironic twist, in the 1950s, Alan Turing, the man who many see as the "father" of modern AI, was persecuted at the hands of the authorities on the basis of his sexual orientation. It is also interesting in this context to tie in Turing's' second perceived objection to the idea that by the year 2000 most people would have no problem with the idea of "thinking machines", with the recent "Gaydar" research (Turing 1950). Turing called this the "Heads in the Sand" objection. This argument went along the lines of

> The consequences of machines thinking would be too dreadful. Let us hope and believe that they cannot do so.

But that was many years ago, you may argue—times and laws and general perceptions have changed in modern liberal democracies. At the very worst, surely the application of this (and similar) technologies constitute a gross invasion of privacy, but nothing more? Perhaps, but in many parts of the world today homosexuality still constitutes a criminal offense, punishable by a variety of sanctions, including death in some cases.

An even greater question arises here. Who actually benefits from research such as this? We may, of course, argue that as an intellectual endeavour it is astonishing to be able to perform this feat, and as such is well worthy of scientific acclaim; it is comparable perhaps to proving some abstruse mathematical theorem, which has no obvious benefit to mankind, but is clearly a significant intellectual achievement. But—there *is* a difference. Proving that abstruse mathematical result in general has no great impact on humankind, except perhaps for that small band of mathematicians who can understand and appreciate the result.

Research such as "Gaydar" however, while being in the same sense a significant intellectual achievement, *does* have the potential to affect a large portion of humanity in a significantly adverse fashion. To quote from the authors themselves:

> ...the predictability of sexual orientation could have serious and even life threatening implications to gay men and women and the society as a whole.

As an illustration let us take another possible scenario. Let's say I spend my entire research career in the investigation and development of a deadly biological/chemical agent that only targets members of the Estonian (or Scottish, or Irish, or whatever) community. Once I have perfected my results, instead of keeping the results secret for the use of whatever military or clandestine organisation to which I might owe my allegiance, I release the results "open source", so to speak, with the declaration "Look, I have, over many difficult years, in semi-secret, managed to develop this deadly agent, which targets only people of Estonian origin. This is to now warn the general public as to the fact that governmental agencies and terrorist organisations may also be capable of harnessing such technology (or, indeed, may already have done so.)"

Was this "ethical research"? To again quote from the author's own article:

> ... we believe that further erosion of privacy is inevitable, and the safety of gay and other minorities who may be ostracized in some cultures hinges on the tolerance of societies and governments.

Another way we can look at research such as this is to ask what potential *good* it can do. To my mind the main (only) positive attribute of this research is that, as it has been published in the public domain, it alerts us all to the possibility of other, far more sinister, research that may currently be being done behind closed and tightly locked doors, by large corporations, world governments, and security agencies. This technology ("Gaydar"), explosive though it potentially might be, can be reproduced reasonably easily, unlike for example, nuclear technology. In fact, the authors take pains to point out that they

> ...used widely available off-the-shelf tools, publicly available data, and methods well known to computer vision practitioners.

At the end of the day, many of these technological "advancements" boil down to algorithms implemented by computer programs, which can be relatively easily shared in a very short space of time using the global Internet. Also. many of the basic algorithms and techniques employed in this research date back to the 1980s and before. The main difference today is the ubiquitous worldwide availability of advanced computer hardware, unheard of in the 1980s. But you may argue, there are reputed to be bomb-making plans of various sorts available on the net (if you know where to look), so is there really any major difference here?

One crucial difference is that the computer code essentially *is*, in many cases, the technology (implemented on some suitable, readily available, hardware platform). In the case of bomb making, for example, possession of plans is just one part of the process. One also needs to get hold of the explosive material, etc. Thus, the potential for the large-scale *implementation* of such highly disruptive, invasive, and potentially harmful technologies is much greater.

13.3   The Internet of Things (IoT)

> You're probably about to lose something precious. Something you can't see. Something you can't touch, taste, or smell and probably don't think about regularly. And yet when it's gone, which I believe it will be soon, you may spend the rest of your life longing for it.
>
> What you're about to lose is your privacy.
>
> Actually, it's worse than that. You aren't just going to lose your privacy, you're going to have to watch the very concept of privacy be rewritten under your nose. That's because while the Internet of Things (IoT) is going to add a lot to our lives, it's probably going to take our privacy in payment, whether you want it to or not.
> —Geoff Webb (5th February 2015). "Say Goodbye to Privacy". WIRED editorial.

The smart power meters that we discussed earlier on in this chapter are just one example of a range of devices that in the past were relatively passive unresponsive devices but which now are interactive devices, typically connected to the Internet, and with advanced sensing, processing, and storage capabilities (Porambage et al. 2016). Undoubtedly this technology has the power to improve people's lives in a multitude of ways including user convenience, allowing for devices such as cookers, fridges and lights to be controlled over the Internet, and for example allowing for a fridge to automatically order supplies such as milk or eggs when supplies run low; potential health benefits in the increasingly powerful and diverse wearable devices that can measure, for example, the wearer's overall activity levels, and blood glucose levels, among other parameters, and efficiencies in areas such as energy and water consumption (Brill 2014).

But these potential benefits come with several downsides—the most important being the potential catastrophic loss of virtually any sense of individual privacy, as evocatively evinced by the quote at the start of this section. In addition, many IoT devices currently operating have shown themselves vulnerable to hacking including, in a recent study the ability to hijack a drone, stealing or even crashing it; the ability to control a child's toy to use it as a spying device or to engage in inappropriate conversations, and the ability to take control of household internet connected cameras (Valente et al. 2019), As the number of IoT connected devices continues to rise (estimated by the International Data Corporation (IDC) to reach 41.6 billion connected devices by the year 2015[1]) exposed vulnerabilities will continue to pose an ever present and growing threat to individual privacy.

13.4   Technology: A Threat to Democracy and Freedom?

AI is a rare case where I think we need to be proactive in regulation instead of reactive—because I think by the time we are reactive in AI regulation it's gonna be too late
    —Elon Musk on AI regulation (in interview), 2017

When advanced technologies such as deep learning are applied to existing democracies there arises the serious issue of control—intelligent control. Norbert Wiener, who we introduced in Chap. 5 said "Cybernetics is the science of information and control, regardless of whether the target of control is a machine or a living organism", and further "Everything is controllable" (Helbing et al. 2017). And in modern advanced societies the raw material that control algorithms require in order to influence and control human behaviour on both an individual and a societal level is plentiful. And here lies the rub. Governments, given the option, will nearly always choose those options which afford them greater surveillance and control over the general population, especially those governments whose position may potentially be disrupted by the will of sizeable segments of their country's population. It is a simple rule of survival. So, the more that these technologies can be "sold" to the general populace— for ostensibly benign purposes of course—the better it suits the lawmakers, keen to maintain their power base.

Let's take a nightmarish scenario. Imagine a world where every citizen was monitored closely and given a rating that changed over time, based on observed behaviour of this individual, including what books people read. This rating, in the range 350–950, governs each citizen's living conditions, such as their access to employment and housing, their ability to obtain a loan, and their ability to travel abroad.

---

[1]https://www.idc.com/getdoc.jsp?containerId=prUS45213219

In the People's Republic of China (PRC) this scenario is not a fiction—it is a reality, today, for certain sections of the population. This Orwellian style surveillance and control system will be mandatory for all citizens of the PRC in 2020 (Helbing et al. 2017). Essentially this system based on the idea of a Citizen Score, assigned to each individual, and which can rise and fall depending on their observed behaviours and which allows for both mass surveillance of the population, and control at both an individual and societal level. While there may well be some advantages to such strict surveillance of individuals, clearly if such a system were to move to democratic countries, there is at least the possibility that the very foundations of their democratic systems could be severely undermined.

Chapter 14
The Future of Society: Choices, Decisions, and Consequences

14.1 A Tipping Point Has Been Reached?

While quite aware of the ever-accelerating rate of progress in the AIE field—(this was, indeed, one of the main motivations for writing this modest text in the first place), I considered it unlikely that a "tipping point" would be reached anytime in the course of my writing this book. By tipping point, I mean that point where a learning intelligent system would be developed with a sufficient level of adaptability that, as a proof of concept it could be potentially successfully applied to a huge range of applications, in which hitherto humans were clearly superior in performance. This tipping point, in my opinion, may well now have been reached as of December 2017. And the name of this tipping point is *AlphaZero*.

14.1.1 From AlphaGo to AlphaZero and MuZero: Glimpses into the Future?

The game of Go, which we originally introduced in Chap. 3 is one of the oldest board games in existence, more than likely developed in China several thousand years ago (Burns 1998). It may be classified as a territorial game where the purpose is to capture territory or the opponent's pieces. Go is also very popular in Japan and Korea. In terms of difficulty level Go has been compared to playing five games of chess simultaneously—one in each corner of the 19 × 19 board, and one in the middle. While the rules may appear simple, it can take many years to master the subtler aspects of the game play. Because of this complexity, up to only a few years ago most experts believed that it would be at least a decade before a computer could successfully challenge the best human players, although some successes had been reported on 'pared-down' Go boards.

© Springer Nature Switzerland AG 2020
M. Eaton, *Computers, People, and Thought*,
https://doi.org/10.1007/978-3-030-55300-5_14

Then in January 2016 a group of researchers at Google's Deepmind facility in London published a paper in the prestigious journal *Nature*, in which they described *AlphaGo*, a Go playing program that used deep reinforcement learning to play the game at a level such that it defeated the European Go champion Fan Hui, in five games out of five (Silver et al. 2016). An enhanced version of AlphaGo (*AlphaGo Lee*) then went on, in March 2016 to beat the Go world champion Lee Sedol in a five-game match in Soeul, Korea. AlphaGo won convincingly in four games out of five in a match, many would consider of equal, if not higher significance to Garry Kasparov's defeat in 1997 to Deep Blue.

The AlphaGo architecture was complex, consisting of two deep neural networks initially primed to emulate Go play by expert human Go players using a large database of recorded human moves using a process of supervised learning. Once a certain level of ability was reached it was then trained on other instances of itself using reinforcement learning techniques. In the words of Thore Graepel, one of the DeepMind creators of AlphaGo (Metz 2016)

> Although we have programmed this machine to play, we have no idea what moves it will come up with. Its moves are an emergent phenomenon from the training. We just create the data sets and the training algorithms. But the moves it then comes up with are out of our hands—and much better than we, as Go players, could come up with.

While few would doubt the significance of AlphaGo as a milestone in the ability of computers to tackle a highly intellectual and tactical challenge better than the best human grandmasters, the ability of AlphaGo to attain this impressive feat was initially bootstrapped by its ability to access, and to subsequently mimic the play of the best human Go players. The subsequent DeepMind development *AlphaGo Zero* required no such initial priming.

In a ground-breaking paper entitled "Mastering the game of Go without human knowledge" published in *Nature* (as was the original AlphaGo article) the authors describe the development and deployment of a new reinforcement learning-based Go playing engine AlphaGo Zero, which emphatically defeated AlphaGo, and more significantly did not require any bootstrapping from human experts. In their own words (Silver et al. 2017)

> Here we introduce an algorithm based solely on reinforcement learning, without human data, guidance or domain knowledge beyond game rules. AlphaGo becomes its own teacher: a neural network is trained to predict AlphaGo's own move selections and also the winner of AlphaGo's games... Starting tabula rasa, our new program AlphaGo Zero achieved super-human performance, winning 100–0 against the previously published, champion-defeating AlphaGo.

AlphaGo Zero is undoubtedly a significant advance in machine intelligence, some would say even more significant than the original AlphaGo, because of the fact that it takes prior human game-playing expertise completely out of the loop.

Then, in December of the same year (2017) an arXiv preprint was published describing a generic algorithm based on AlphaGo Zero called simply *AlphaZero* which was applied not just to Go, but also to chess and shogi (Japanese chess). The final version of this article was subsequently published in the prestigious journal *Science* (Silver et al. 2018).

Like AlphaGo Zero, AlphaZero uses at its core a search algorithm called Monte-Carlo tree search (MCTS). At its most basic level this algorithm essentially involves choosing a move from a current board position by looking at each possible move from this position and simulating playing this move, and subsequently playing random moves by each player until a final result is reached. This process is followed many times for each possible board move and the average winning rate for each move is computed. Finally, the move with the highest winning rate is chosen as the actual move to be taken. AlphaZero also keeps a visit count, mover probability, and value (based on previous simulated moves) and uses these to guide move selection rather than purely random move taking, as in the basic version of MCTS described above.

The power of AlphaZero is evident from the fact that, following training, it was capable of defeating both AlphaGo Lee and AlphaGo Zero in the game of Go. It also defeated *Stockfish*, one of the strongest chess engines based on the results of the 2016 *Top Chess Engine Competition* (TCEC), the world championship of chess engines, and, in addition defeated *Elmo*, the 2017 world champion shogi program according to the *Computer Shogi Association* (CSA). Interestingly, both Stockfish and Elmo utilised highly optimised and specifically adapted alpha-beta algorithms at their core, as discussed in Chap. 7. More recently Facebook published an open-source re-implementation of the AlphaZero algorithm called *ELF (Extensive Lightweight, and Flexible) OpenGo* (Tian et al. 2019). This paper also analyses how the algorithm's behaviour alters during training. They also released an OpenGo public Windows binary, which runs on either a CUDA-enabled graphics processing unit (GPU), or (more slowly) on a straightforward CPU.

Then, in November 2019 the latest edition of the AlphaGo series was released—MuZero. Based broadly on the AlphaZero architecture, in addition to achieving superhuman performance in chess, shogi, and Go; indeed, in fact surpassing AlphaZero in Go, it also achieved a new state of the art in playing a suite of 57 Atari games that is the standard test suite for AI video game performance testing. In addition, and unlike AlphaZero, MuZero only checks for legal moves at the root of the current search tree and does not check for the legality of moves for the remainder of the search tree. Related to this, it does also does not check for episode termination within the search tree, relying instead always on the value predicted to be returned.

14.2 The "Mechanisation" of Human Endeavour

While, of course, there have been questions and fears about the potential of AI from well before the term "Artificial Intelligence" was bandied about in the mid-1950s, I began to have vague stirrings of serious unease about the whole area of technological "advancements" in the early 2000s, to the extent that I felt it incumbent on me to include a short section on his topic as part of my 2008 article "Evolving humanoids: using artificial evolution as an aid in the design of humanoid robots" (Eaton 2008). I include a brief quote from this article below.

> As AI researchers strive, on one hand, to recreate human-like intelligence in machine form on the other hand people are being coerced and cajoled into acting and thinking in an increasingly mechanised fashion. While machines may indeed one day achieve human-level intelligence (or beyond), however we define this, the opposite will never be the case. Indeed, the gap can only get wider as machinery and its computing ability becomes ever more sophisticated. . .Rather than blindly embracing future technologies for technologies' sake we should be more critical as to their potential future benefits and drawbacks for humanity as a whole.

I was not really aware of how prophetic these words might be when I wrote them, over 10 years ago now. People are being asked to work more like machines meeting exact targets, with much of the joy being taken out of the workplace environment. We can identify a number of areas in which machines clearly have advantages over humans—of course, many of these areas will also have been recognized from the dawn of the industrial revolution, and well before.

- Machines don't mind ever-increasing bodies of rules and regulations—this is how machines operate best, the more rules the better
- Machines have no privacy concerns
- Machines are "happy" to be on call/available 24/7 (subject to maintenance)
- Machines have no concern in having their performance "evaluated" on a highly frequent basis

Decisions need to be made about the future—one decision critical at the present time is: who/what would you rather deal with in everyday interactions—a machine, or a human? Unfortunately, in some recent research initiatives this question has been answered—and the answer has not favoured humans. The thing is—*humans are not machines*. So, encroaching workplace practices aimed at making humans work more like machines (aren't machines perceived as being more effective than humans in so many walks of life nowadays, and increasingly so?), may in the end turn out to be highly counterproductive, in the long run.

Digital computation, at its core, as we have seen, reduces to a simple mechanic— *switches driving switches*. Do we, as humans, wish to be reduced to thinking of ourselves as such? I think not. Remember: *we are not machines*.

14.2.1   Perils of Advanced Technologies/Looking to the Future

We can draw an analogy (and there are many) between the development of computational prowess by mankind, and the, in certain senses associated, development of atomic energy and of nuclear bombs. Both of these distinct areas of endeavour were "discovered" by men (and some women) of exceptional intellect. Both areas of research and endeavour also have the ability to deliver extraordinary benefits to mankind—and also extraordinary ills; including, in both cases the elimination of humanity.

It is clear also that one further avenue of research at this time is of particular concern; that is the whole area of biotechnological research involving novel techniques allowing for the precise manipulation of genomic material in animals and also in humans, with all that this implies.

Humanity would do well to draw back from our self-destructive spiral fuelled by advancing technological advances, to gaze back at the fundamental issues of importance to all of humankind. These encroaching technologies can be used for incredibly destructive purposes, or, to flip the switch, for the gradual betterment of all of humankind including, even, potentially the creation of a helpful new "race" of robotic semi-citizens.

Wars and destructive behaviour are, many times, born and fermented from perceived inequalities. We now possess, with the beneficial use of modern technologies, at the forefront of which is the AI/robotics field, the ability to gradually erode these inequalities worldwide. Is this alone enough to make people happy? Perhaps not. Maybe there is a streak in humanity that will always strive to conquer, to be better than our neighbours.

But, at least, with sufficient constraints and controls over these emerged and emerging technologies, mankind's baser instincts can, to a large extent, be limited. We have seen this approach applied successfully (to date) to the proliferation of nuclear weapons worldwide. The same should be made to apply to new and advancing technologies. The alternative dystopia must be avoided at all costs. Important decisions and policy directives should be made on the basis of long-term benefits to humanity as a whole, and not on corporate greed, or on primarily financial or otherwise selfish motives, likely to benefit only the elite few.

The real danger, certainly in the medium term, may be not so much what technology can do (or will be able to do in the future), but how we (humans) use this technology. The potentiality for nuclear power and nuclear weapons has always existed; it is the manner of their deployment that is the issue. Similarity it might be argued that, from the early days of the universe, the potentiality for non-biological life forms and "intelligence" has always existed—however it required an "intervening" phase of biological intelligence to facilitate such intelligence.

However, this then brings into question what exactly we mean by "artificial". Are biological entities, in one sense, not artefacts in their own right, facilitated by a process of natural evolution? So, for what reason do we say that biological entities ("old intelligence") cannot coexist peacefully and in harmony in the future with our "artificial" (new intelligence) colleagues?

14.3   Responsible AI

> A new and more ambitious form of governance of AI systems is a most pressing need. One
> that ensures and monitors the chain of responsibility across all the actors. This is required to
> ensure that the advance of AI technology is aligned with societal good and human well-
> being.
>   — Virginia Dignum *Responsible Artificial Intelligence* (2019)

Virginia Dignum, author of the above quote, in her book *Responsible Artificial Intelligence* (Dignum 2019) emphasises the importance of the participation and of the active inclusion of all strata of society in the making of important decisions as to the direction that the development of advanced AI takes.

To this end she stresses the importance of the education of the general public as to what AI is and as to its potential impact on their lives. She also emphasises the importance of the education of AI researchers as to the potential ethical and societal implications of their work. Both of these foci lie at the core of my rationale and purpose in writing this book.

In the context of responsibility this can be seen to spread not just among AI researchers and developers but also over the users of advanced AI technology and over educators, and also over governments who (should) have the power to ban or to restrict certain disruptive technologies if they do not judge them as beneficial to society as a whole. As Dignum puts it

> Responsible AI is more than ticking ethical boxes in a report or developing add-on features
> or switch-on buttons in AI systems. Rather, it is the development of intelligent systems
> according to fundamental human principles and values. Responsibility is about ensuring that
> results are beneficial for many instead of a source of revenue for a few.

Dignum stresses three core principles in the development of responsible AI—Accountability, Responsibility, and Transparency (ART). Central to these principles is the requirement that the AI system be able to *explain* clearly to the user *why* it is taking certain decisions, and also that the overall rationale for the decisions made by the system in the first place are made clear in the context of prevailing societal standards and values.

For a good introduction to the field of *Explainable Artificial Intelligence* (XAI), see the article by Arrieta et al. (2020) which provides the straightforward definition of the field:

> Given an audience, an explainable Artificial Intelligence is one that produces details or
> reasons to make its functioning clear or easy to understand.

14.3.1   A Code of Ethics for AI?

In her 2017 book *Towards a Code of Ethics for AI*, Paula Boddington, a philosopher who has published a significant amount of work in the area of medical ethics, also emphasises the importance of transparency in AI operation, together with transparency and discourse in the development of any code of ethics specific to any organisation or country (Boddington 2017). A notable point is that this book was funded by the Future of Life Institute (FLI) mentioned in Chap. 9, in relation to AI safety funding, as part of the project "Towards a code of ethics for AI research".

Interestingly, Boddington makes the point in her book that, counterintuitively, the development of a code of ethics could actually make the situation worse, referencing Milgram's classic *Obedience to Authority* experiments where otherwise upstanding individuals were coaxed into administrating what they believed to be large and painful shocks to complete strangers (Milgram and van Gasteren 1974). In her own words

> expressing a moral sentiment may in some circumstances decrease the likelihood of behaviour that follows one's conscience.

Boddington also cautions against the notion of stressing intelligence as humanity's main distinguishing feature. Indeed, scientists believe that several other species on Earth, including a number of the higher primates, elephants, and dolphins are very likely possessed of intellectual abilities (however we define these) that are not necessarily so far removed from those of humans. Quoting from Aristotle, she cites empathy, social skills, and "quite possibly" religion as other core factors in the essence of our shared humanity, as evidenced in the following quote (Boddington 2017).

> If we see AI as human progress, if we are concerned about the ethics of AI, we must guard against a simplified attention to bare intelligence and to idealised, isolated individual agency.

14.3.2 A Modest Proposal

In view of (and even despite of) the previous discussions a case could certainly be reasonably made that certain restrictions should be placed on the widespread availability and/or deployment of potential future disruptive technologies based on the democratic decision of a systematic random survey of ordinary individuals (not politicians, bureaucrats, or businessmen) in this regard. The following proposal does not forbid research in these controversial areas (although separate legislation may be required, and, indeed, enacted in many jurisdictions regarding such technologies as cloning, etc.), but limits exposure of the general public to technologies that might be considered to be very harmful.

This proposal might in a sense be considered "big brother" in its nature and in its tone but it attempts to include the critical involvement of ordinary citizens in each important pronouncement. An outline of this proposal is contained in the next section.

Bruno Frey also makes the case for the active participation of citizens regarding the adoption of the adoption of future technologies and advocates the holding of popular referenda that will allow direct participation in the decision-making process following an initial narrowing down of the options available (Helbing et al. 2017).

He also makes the interesting proposal for the adoption of each official state body of an "advocatus diaboli", whose job it is to produce counter-arguments to each proposal made, thus fuelling a search for alternative proposals and reducing the probability of decisions being made based on "political correctness" considerations.

14.3.3   On the Adoption/Prohibition of Advanced Potentially Disruptive Technologies: A Provisional Manifesto

Based on recent technological advances we need a radical way of reorganising human society for the betterment of all, while still recognising the exceptional achievements of the few.

Given: that the decision to adopt advanced disruptive technologies should not be left in the hands of governments alone, or in the hands of the scientists or engineers involved in the development/deployment of such technologies, given the major vested interests of both of these parties, *we advocate* the establishment of a citizens convocation on technology to provide an impartial (insofar as is possible) assessment of the risks and benefits of said technologies.

Proposal: That each sovereign country convene a "citizens' convocation on technology" comprising a minimum of 300 citizens of that country or 0.005% of the population, whichever is the greater number, divided into three equal groups, chosen using systematic random-sampling techniques from the socio-economic fabric of that country. The purpose of the convocation is to make decisions regarding the adoption or otherwise of technological advancements following presentations to each grouping independently of the details of said technologies by the developers and experts in the particular technology and following arguments both in favour and against such technologies by eminent speakers. In the event of disagreement between the three groups a simple majority will suffice.

Decisions available to the citizens' convocation include:

A. Allow said technological advancement(s) to be adopted freely and without curtailment.

B. Allow for the adoption or rejection by individual citizens of the "advancement" based on a clear exposition of the risks and benefits associated with said technologies (e.g. "smart" meters).

C. Prohibit the development, use or deployment of said technologies within the borders of the sovereign state. If this results in, for example, restriction of Internet access to certain facilities available beyond the country's borders this restriction shall be rigorously enforced.

Given the adoption of resolution (C) by the citizens' convocation, a minimum of 7 years shall elapse before the re-consideration of said technology by the convocation.

14.4 Conclusions

The computer, as we have seen, has heralded and facilitated wondrous advances in humans' ability to manage, process, and store information, leading to a host of beneficial technological innovations. (Where would we be without the microprocessor-controlled clothes iron?). However the computer has also heralded and facilitated some of the most damaging and potentially destructive advances in the sciences and in technology, including (from its earliest days) facilitating the design and development of nuclear weapons, which, to this day, threaten the very existence of humankind, now and into the foreseeable future.

Of course, there are those who would argue that also, today, the use of advanced supercomputers for simulation purposes reduces the "requirement" for the testing of real nuclear weaponry, with the associated benefits to our environment. We might make a (rather crude) analogy with guns and the weapons industry in general. In the United States currently statistics show that about the same number of people die from firearm deaths as die on the roads. Yet a large (and vocal) portion of the American population are in favour of the right to bear arms, and the US National Rifle Association (NRA) quotes the mantra "Guns don't kill people; people kill people". Yes, indeed. But is anybody seriously arguing that if all of the guns were taken out of American civil society, the domestic (and overall) death rate would not plummet?

No one (except the most fanatical) is suggesting that we should do away with all computer-related and other advanced technologies completely. However, a carefully considered approach as to which technology or groups of technologies may usefully serve mankind into the future, and which should be restricted, or even curbed entirely, is perhaps, well overdue. Technology, of which the modern digital computer is the supreme example today, is a double-edged sword. Let us take all measures possible to sharpen that edge which bodes well for humankind, but also to dull that which acts for evil. Our very future existence may now depend on it.

So—you might say—hollow platitudes, what concrete advice can be offered?

Well, perhaps the next time you are given the option of a "convenient" automated checkout in a shop or a cafe, maybe go for the human option instead (it might even save you some time). If you are in a cafe or a bar, on your own, with a little time to spare, instead of spending the entire time "interacting" with your smartphone, perhaps spend a few minutes in idle conversation with a real human. In the words of Max Tegmark (2017):

> Do you want to be someone who interrupts all their conversations by checking their smartphone, or someone who feels empowered by using technology in a planned and deliberate way? Do you want to own your technology or do you want your technology to own you... Our future isn't written in stone and just waiting to happen to us—it's ours to create. Let's create an inspiring one together!

And *question*. When a new technological "advancement" is proposed, perhaps allied to Facebook, Twitter, etc., question whether this is really going to improve *my* quality of life—or am I just adopting this technology in order to show how "technologically savvy" I am, and to hell with any long term consequences that may accrue. There is a high likelihood that in virtually all of the abilities and competencies at which humans are currently better than machines we will be overtaken, if not in the next decade, almost certainly in the next 30 years or so. However, as humans, we will always, in many senses, prefer interaction with real humans—so our working roles in society are not completely extinct!

However, if there is no respect paid to essential human dignity then, of course, we must all bow before the mighty machine. After all, computers/robots are stronger than us, faster than us, and in the near future may well be our intellectual equals/superiors in virtually every area of human endeavour. In short, we need, in a sense, to grow up and to respect ourselves, and our innate humanity. Human culture over machine "culture". There is more than enough wealth in this world to go around, it seems, if distributed in an equitable fashion. In the (relatively) near future we will have an army of robotic helpers available, allowing for the eradication of the vast majority of so-called D^3 (dirty, dangerous, or dull) work.

On the other hand, if mankind persists (as is unfortunately currently happening in many instances) in the application of exponentially increasing technological advances to, in the evocative words of the late bridge-builder, entrepreneur, and eminent robotics researcher Takashi Gomi (now, sadly, passed away), "the dark side", then future outcomes are far more uncertain.

Appendix A
What Is the Most Efficient Number Base?

A.1 Comparison of Different Number Bases

We will now conduct a brief analysis of several different number bases in terms of representational efficiency and cost, assuming a polynomial cost function. We assume here a representational cost of a number as $f(r).l$ where r is the radix of the number and l is the length in digits, and furthermore we assume $f(r)$ is of polynomial form, that is the cost reduces to $r^x.l$, where x is a positive value.

In the simplest case, taking $x=1$ we have $R(n)=r.l$: that is the representational cost of a number n is the product of the radix and the length of the number we have the following analysis, as demonstrated in Table A.1.

A.2 Comparison of a Single Duodecimal Digit to Binary, Ternary, Quaternary, and Base 6 (Senary) Taking a Fixed Representational Cost of 12

A.2.1 Case Where x=1

For binary, this allows for a six-bit number ranging in value from 000000 to 111111—64 combinations.

For ternary, this allows for a four-trit number ranging in value from 0000 to 1111—81 combinations.

For quaternary, this allows for three quaternary digits, ranging in value from 000 to 333—64 combinations, as in binary.

For senary (base 6) this allows for two senary digits, ranging in value from 00 to 55—36 combinations.

For duodecimal, this allows for a single digit duodecimal number ranging from 0 to B—12 combinations (See Fig. A.1 for a comparison of the number of

© Springer Nature Switzerland AG 2020
M. Eaton, *Computers, People, and Thought*,
https://doi.org/10.1007/978-3-030-55300-5

Table A.1 Number of different states available in different radices for different values of x≤1

Base	x=1	x=0	x=0.5 (min/max states)	x=0.9 (digits/min/max states)
Binary—B	64	4096	256/512	6.43/64/128
Ternary—T	81	531,441	729/2187	4.48/81/243
Quaternary—Q	64	16,777,216	4096	3.45/64/256
Senary—S	36	2,176,782,336	1296/7776	2.39/36/216
Duodecimal—D	12	$8,916,100,448 \times 10^{12}$	1728/20,736	1.28/12/144

Fig. A.1 A comparison of the number of combinations possible for a single duodecimal digit in binary, ternary, quaternary, and senary (base 6) bases

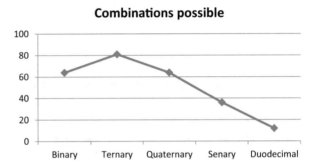

combinations possible for a duodecimal digit in the binary, ternary, quaternary, and senary bases).

In this case (x=1) for polynomial functions of $f(r)$, ternary is clearly the winning radix with 81 combinations (which can, of course, represent numbers, letters, pixels, etc.) followed by binary and quaternary which tie for second place with 64 combinations each, then senary with 36 combinations, and finally duodecimal with just 12 combinations (the digits 0 to B).

A.2.2 For x≤1

For the trivial case where x=0, that is the representational cost does not vary relative to the radix used (an unlikely scenario in real-world scenarios), duodecimal (the highest radix) is the winner, and binary is the worst radix by this reckoning; indeed clearly the higher the number base used, the greater the efficiency by this reckoning. Table A.1 lists the number of different states available for binary(B), ternary(T), quaternary(Q), senary (S), and duodecimal (D) for several different values of x≤1.

A.3   Values of x>1 Taking a Fixed Representational Cost of 36

Taking x=2 we have the following:

For binary, this allows for an eight-bit number ($36/2^2$), ranging in value from 00000000 to 11111111—256 combinations.

For ternary, this allows for a four-trit number ($36/3^2$), ranging in value from 0000 to 1111—81 combinations.

For senary (base 6) this allows for one senary digit ($36/6^2$) ranging in value from 0 to 5—six combinations.

Taking x=3 we have the following:

For binary this allows for a four-bit number ($36/2^3$)—16 combinations

For ternary we have a 1/2 trit number ($36/3^3$)—three or nine combinations.

It seems clear from this rudimentary analysis that for values of x substantially greater than 1, the lower the number base the greater the representational efficiency—i.e., in this case binary is the most efficient system.

A.4 Conclusions

So, assuming we do not have at our disposal some magical material which allows us to store ever-increasing numbers of states at a single location at no extra cost, the ternary base fares well, in many cases better than binary. Of course, other factors also affect our choice of number base, and it can be argued that the balanced representation of numeric data is more elegant and allows for more efficient circuitry: only odd number bases can have this property, which excludes binary. Taking all of these factors into account ternary would appear to be, in very many respects, the most desirable elegant and efficient system.

Appendix B
Ternary State Machine Design

B.1 Design of a Two-Trit Balanced Ternary Counter

To illustrate the application of the algorithmic state machine design process to the ternary case we will consider two examples, a two-trit balanced ternary counter with nine states in total, and a similar two-trit counter with input. Firstly, we will look at the excitation tables for two forms of ternary memory elements which we will call the *delay* (D), and *toggle/rotate* (T/R) operators. These operators may be given different labels in the ternary computation literature as there are little decided nomenclature standards in this area. As mentioned earlier, rather than the term flip-flop which is sometimes used to denote a memory element in the binary domain, we now use the term flip-flap-flop (FFF) to highlight the three-state nature of these devices.

Initially we just look at the operations engendered by these two operators, as demonstrated in Tables B.1 and B.2:

And their excitation tables, as shown in Table B.3:

The ASM chart for the two-trit counter is given in Fig. B.1. This counter counts from -4 to $+4$ sign balanced ternary notation, so we go from $- -$ to $+ +$, with nine states in total.

Table B.1 Delay ternary FFF

Input	Action	Comment
−	−	Stay
0	0	Stay
+	+	Stay

Table B.2 Toggle/rotate ternary FFF

Input	Action	Comment
−	↑	Rotate left
0	−	Stay
+	↓	Rotate right

© Springer Nature Switzerland AG 2020
M. Eaton, *Computers, People, and Thought*,
https://doi.org/10.1007/978-3-030-55300-5

Table B.3 Delay and rotate ternary excitation tables

P	N	Delay	Toggle/rotate
−	−	−	0
−	0	0	+
−	+	+	−
0	−	−	−
0	0	0	0
0	+	+	+
+	−	−	+
+	0	0	−
+	+	+	0

Fig. B.1 ASM chart for a nine-state ternary counter

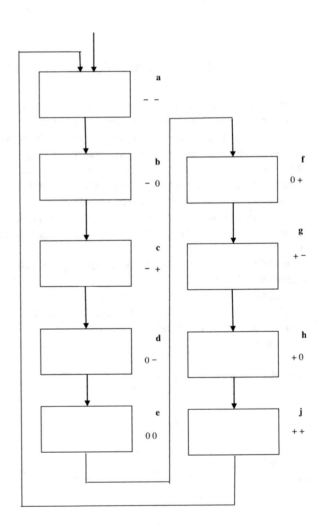

Table B.4 Next state table for the two-trit nine-state ternary counter

State name	Present state	Next state	T/R B	T/R A	DB	DA
a	– –	– 0	0	+	–	0
b	– 0	– +	0	+	–	+
c	– +	0 –	+	+	0	–
d	0 –	0 0	0	+	0	0
e	0 0	0 +	0	+	0	+
f	0 +	+ –	+	+	+	–
g	+ –	+ 0	0	+	+	0
h	+ 0	+ +	0	+	+	+
j	+ +	– –	+	+	–	–

Table B.5 K-map simplification of nonary counter using ternary T/R FFF technology

T/R A input

A/B	−1	0	+1
−1	+1	+1	+1
0	+1	+1	+1
+1	+1	+1	+1

T/R B input

A/B	−1	0	+1
−1	0	0	+1
0	0	0	+1
+1	0	0	+1

Now to look at the next state table for a nine-state −4 to +4 ternary synchronous counter, using T/R and D FFFs, as illustrated in Table B.4.

This gives the K map for the T/R FFF in Table B.5.

This gives rise to the simplified equations (using R rather than T/R)

$$RA = +1$$

$$RB = \overrightarrow{A}$$

with the corresponding circuit diagram shown in Fig. B.2:

We can also generate the K map for the D FFF, as shown in Table B.6:

This gives rise to the simplified equations

$$DA = \underline{A} + \left[\overrightarrow{A}\right]$$

$$DB = \left[\overleftarrow{\underline{A}.\overleftarrow{B}}\right] + \left[\underline{A}.\overrightarrow{B}\right] + \overrightarrow{A}.\underline{B} + \left[\overrightarrow{A}.\overrightarrow{B}\right]$$

Fig. B.2 Nine-state
(nonary) counter using T/R
flip-flap-flop

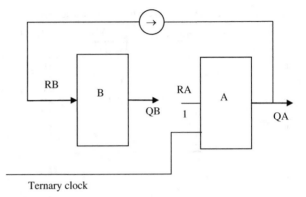

Ternary clock

Table B.6 K-map simplification of nonary counter using ternary D FFF technology

DA input

A/B	−1	0	+1
−1	0	+1	−1
0	0	+1	−1
+1	0	+1	−1

DB input

A/B	−1	0	+1
−1	−1	−1	0
0	0	0	+1
+1	+1	+1	−1

with quite straightforward ternary logic circuit implementation, which the interested
reader may wish to generate as an exercise.

B.2 Ternary State Machine Design: With Input

Let us now take the same example as in the previous case—a nonary ternary counter,
but this time add in an input D; the ASM chart is as given in Fig. B.3.

We can now generate the corresponding K-maps using the variable-entered-map
(VEM) approach, adapted to the ternary case, as shown in Table B.7. D here is the
ternary map entered variable, taking the one of the values −1, 0, or +1.

This gives rise to the simplified equations (shortening to R rather than T/R)

$$RA = \vec{A} + \overleftarrow{A} + \vec{B} + \overleftarrow{B} + \underline{A}.\underline{B}.\left(\left[\vec{D} \right] + D \right)$$

$$RB = \vec{A} + \underline{A}.\underline{B}.\left[\vec{D} \right]$$

Fig. B.3 ASM chart for a
nine-state ternary counter
with input

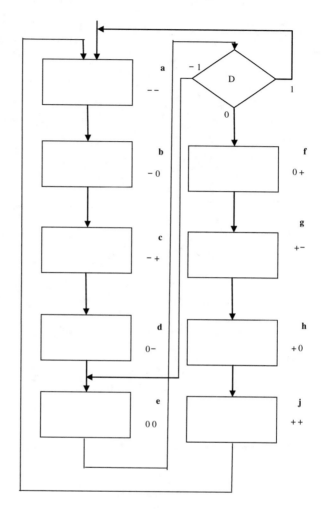

Table B.7 K-map simplification of nonary counter with input using ternary T/R FFF technology

T/R A input   T/R B input

A/B	−1	0	+1
−1	+1	+1	+1
0	+1	$[D]+D$	+1
+1	+1	+1	+1

A/B	−1	0	+1
−1	0	0	+1
0	0	$[\vec{D}]$	+1
+1	0	0	+1

References

Aagaard, J. (2015). Media multitasking, attention, and distraction: a critical discussion. Phenomenology and the Cognitive Sciences, 14(4), 885-896.

Abbott, R., & Bogenschneider, B. (2018). Should robots pay taxes: Tax policy in the age of automation. Harv. L. & Pol'y Rev., 12, 145.

Ackermann, R. R., Mackay, A., & Arnold, M. L. (2016). The hybrid origin of "modern" humans. Evolutionary Biology, 43(1), 1-11.

Adler, I. (1961). *Thinking machines, a layman's introduction to logic, Boolean algebra, and computers*. New York: John Day Co.

Aiken, M. (2017). *The Cyber Effect: A Pioneering Cyberpsychologist Explains how Human Behavior Changes Online*. Spiegel & Grau.

Aleksander I.., & Burnett, P. (1987). *Thinking Machines; The Search for Artificial Intelligence*. Oxford University Press

Amarel, S. (1968). On representations of problems of reasoning about actions. Machine intelligence, 3(3), 131-171.

Amari, S. I. (1977). Neural theory of association and concept-formation. Biological cybernetics, 26 (3), 175-185.

Arakawa, T., and Fukuda, T. (1996). Natural motion trajectory generation of biped locomotion robot using genetic algorithm through energy optimization. In: Systems, Man, and Cybernetics, 1996., IEEE International Conference on (Vol. 2, pp. 1495–1500). IEEE.

Arkin, R. (2009). *Governing lethal behavior in autonomous robots*. CRC Press.

Arkin, R. (1998). *Behavior-based robotics*. MIT press.

Arkin, R. (2007). Governing Lethal Behavior: Embedding Ethics in a Hybrid Deliberative/Hybrid Robot Architecture, Report GIT-GVU-07-11, Atlanta, GA: Georgia Institute of Technology's GVU Center.

Arrieta, A. B., Díaz-Rodríguez, N., Del Ser, J., Bennetot, A., Tabik, S., Barbado, A., ... & Chatila, R. (2020). Explainable Artificial Intelligence (XAI): Concepts, taxonomies, opportunities and challenges toward responsible AI. Information Fusion, 58, 82-115.

Asimov. I. (1950). *I, Robot*. Gnome Press.

Babbage, C. (1864). *Passages from the Life of a Philosopher*. John Murray.

Bachrach, Y., Graepel, T., Kohli, P., Kosinski, M., & Stillwell, D. (2014, May). Your digital image: factors behind demographic and psychometric predictions from social network profiles. In Proceedings of the 2014 International conference on Autonomous agents and multi-agent systems (pp. 1649-1650). International Foundation for Autonomous Agents and Multiagent Systems.

Barto, A. G., Sutton, R. S., & Anderson, C. W. (1983). Neuronlike adaptive elements that can solve difficult learning control problems. IEEE transactions on systems, man, and cybernetics, (5), 834-846.

Beckel, C., Sadamori, L., Staake, T., & Santini, S. (2014). Revealing household characteristics from smart meter data. Energy, 78, 397-410.

Beland, L. P., & Murphy, R. (2016). Ill communication: technology, distraction & student performance. Labour Economics, 41, 61-76.

Berkeley, E. C. (1949). *Giant Brains or Machines That Think.* Wiley & Sons.

Berkeley, E. C. (1961). *Giant Brains,* 2nd Edition, Wiley & Sons.

Berlekamp, E. R., Conway, J. H., & Guy, R. K. (2001). *Winning Ways for your Mathematical Plays.* 4 vols. (2nd ed.). A K Peters Ltd.

Blake D. V., & Uttley A. M. (eds.) (1959). *Proceedings of the Symposium on Mechanisation of Thought Processes,* Vols. 1 and 2, London: Her Majesty's Stationary Office, 1959.

Boddington, P. (2017). Towards a code of ethics for artificial intelligence. Springer International Publishing.

Bongard, J. C. (2013). Evolutionary robotics. Communications of the ACM, 56(8), 74–83.

Boole, G. (1847). *The mathematical analysis of logic.* Philosophical Library.

Boole, G. (1854). *An investigation of the laws of thought: on which are founded the mathematical theories of logic and probabilities.* Dover Publications.

Bostrom, N. (2014). *Superintelligence: Paths, dangers, strategies.* OUP Oxford.

Bourg, D. M., & Seemann, G. (2004). AI for game developers. O'Reilly Media, Inc.

Brabazon, A., O'Neill, M., & McGarraghy, S. (2015). *Natural computing algorithms.* Berlin: Springer.

Braitenberg, V. (1986). *Vehicles: Experiments in synthetic psychology.* MIT press.

Brill, J. (2014). The internet of things: Building trust and maximizing benefits through consumer control. Fordham L. Rev., 83, 205.

Brockman, J. (2015). *What to think about machines that think: today's leading thinkers on the Age of Machine Intelligence.* HarperCollins Publishers.

Brooks, R. A. (1991). Intelligence without representation. Artificial intelligence, 47(1-3), 139-159.

Brooks, R.A. (2017). Robotic cars won't understand us, and we won't cut them much slack. IEEE Spectrum, 54(8), 34-51.

Buckland, M. (2005). *Programming Game AI by Example.* Jones & Bartlett Learning.

Burns, B. (1998). *The Encyclopedia Of Games: Rules and Strategies for More Than 250 Indoor and Outdoor Games, from Darts to Backgammon.* Metro Books.

Campbell, M., Hoane, A. J., & Hsu, F. H. (2002). Deep blue. Artificial intelligence, 134(1-2), 57-83.

Carr, N. (2014). *The glass cage: How our computers are changing us.* WW Norton & Company.

Carrel, A. (1935). *Man, the Unknown.* New York: Harper & Brothers

Carter, S. P., Greenberg, K., & Walker, M. S. (2017). The impact of computer usage on academic performance: Evidence from a randomized trial at the United States Military Academy. Economics of Education Review, 56, 118-132.

Christian, B. (2011). *The most human human.* Viking, Great Britain.

Chu, S. C., Tsai, P. W., & Pan, J. S. (2006). Cat swarm optimization. In Pacific Rim International Conference on Artificial Intelligence (pp. 854-858). Springer, Berlin, Heidelberg.

Colomi, A., Dorigo, M., & Maniezzo, V. (1991). Distributed optimization by ant colonies. Proc. of 1st European Conf. on Artificial Life. Paris France: Elsevier.

Cunningham, G. W. (1910). Thought and reality in Hegel's system (No. 8). Longmans, Green.

Darwin, C. (1859). *On the origin of species.* London: Murray.

Dawkins, R. (1986). *The blind watchmaker: Why the evidence of evolution reveals a universe without design.* New York: WW Norton and Company.

de Garis, H. (1990a). Genetic programming: Building nanobrains with genetically programmed neural network modules. In: Neural Networks, 1990 IJCNN International Joint Conference on (pp. 511–516). IEEE.

de Garis, H. (1990b). Genetic Programming: Building Artificial Nervous Systems Using Geneti-cally Programmed Neural Network Modules (1990), Proceedings of the 7th International Conference on Machine Learning, 1990, pp.132–139.

de Garis, H. (1990c). Genetic programming: Evolution of time dependent neural network modules which teach a pair of stick legs to walk. In: Proceedings of the 9th European conference on artificial intelligence (pp. 204–206). Stockholm, Sweden.

de Garis, H. (2005). *The Artilect War: Cosmists Vs. Terrans: A bitter controversy concerning whether humanity should build godlike massively intelligent machines.* Palm Springs, CA: ETC Publications. ISBN 0-88280-154-6.

Deb, K., Pratap, A., Agarwal, S., and Meyarivan, T. (2002). A fast and elitist multiobjective ge-netic algorithm: NSGA-II. Evolutionary Computation, IEEE Transactions on, 6(2), 182–197.

Deb, S., Fong, S., & Tian, Z. (2015, October). Elephant search algorithm for optimization problems. In Digital Information Management (ICDIM), 2015 Tenth International Conference on (pp. 249-255). IEEE.

Descartes, R. (1837). *Discourse on Method* [Translated from French: Discours de la methode].

Devasena, C. L., Sumathi, T., & Hemalatha, M. (2011). An experiential survey on image mining tools, techniques and applications. In International Journal on Computer Science and Engineer-ing (IJCSE).

Dignum, V. (2019). Responsible Artificial Intelligence: How to Develop and Use AI in a Respon-sible Way. Springer International Publishing.

Domingos, P. (2015). *The master algorithm: How the quest for the ultimate learning machine will remake our world.* Basic Books.

Donovan, T., & Garriott, R. (2010). *Replay: The history of video games.* Lewes: Yellow Ant.

Dormehl L. (2016) *Thinking Machines: The inside story of Artificial Intelligence and our race to build the future.* Ebury Publishing.

Dorn, S. D. (2015). Digital Health: Hope, Hype, and Amara's Law. Gastroenterology, 149(3), 516-520.

Dylan, B. (2004). *Chronicles.* Simon & Schuster.

Eaton, M. (2008). Evolving humanoids: Using artificial evolution as an aid in the design of humanoid robots. In Frontiers in Evolutionary Robotics. InTech.

Eaton, M. (2012). Design and construction of a balanced ternary ALU with potential future cybernetic intelligent systems applications. In Cybernetic Intelligent Systems (CIS), 2012 IEEE 11th International Conference on (pp. 30-35). IEEE.

Eaton, M. (2015). *Evolutionary Humanoid Robotics.* Springer Berlin Heidelberg.

Eaton, M. (2016). Bridging the Reality Gap A Dual Simulator Approach to the Evolution of Whole-Body Motion for the Nao Humanoid Robot. In IJCCI (ECTA) (pp. 186-192).

Eccles, J. C. (1973). *The understanding of the brain.* McGraw-Hill.

Engelbrecht, A. P. (2002) *Computational Intelligence: An Introduction.* John Wily & Sons.

Erol, O. K., & Eksin, I. (2006). A new optimization method: big bang–big crunch. Advances in Engineering Software, 37(2), 106-111.

Farley, B.G., & Clark, W.A. (1954). Simulation of self-organizing systems by digital computer. Transactions of the IRE Professional Group on Information Theory, 4(4), 76-84.

Feigenbaum, E. A., & Feldman, J. (1963). *Computers and thought.* McGraw-Hill, New York.

Fisher, D. H. (1987). Knowledge acquisition via incremental conceptual clustering. Machine learning, 2(2), 139-172.

Fleischman, W. M. (2015). Just say "no!" to lethal autonomous robotic weapons. Journal of Information, Communication and Ethics in Society, 13(3/4), 299-313.

Floreano, D., & Mondada, F. (1996). Evolution of homing navigation in a real mobile robot. IEEE Transactions on Systems, Man, and Cybernetics, Part B: Cybernetics, 26(3), 396–407.

Flynn A, Brooks R. (1989). "Battling reality", AI Memo No. 1148, MIT AI Laboratory, Cambridge MA, July 1989.

Fogel, L. J., Owens, A. J., and Walsh, M. J. (1966). *Artificial Intelligence Through Simulated Evolution.* New York: Wiley, 1966.

Ford, M. (2015). *The rise of the robots: Technology and the threat of mass unemployment.* Oneworld Publications.

French, C.S. (1980) *Computer Science: an instructional manual.* D.P Publications.

Friedberg, R. M. (1958). A learning machine: Part I. IBM Journal of Research and Development, 2 (1), 2-13.

Friedberg, R. M., Dunham, B and North, J. H. (1959). A learning machine: Part II. IBM Journal of Research and Development, 3:282-287.

Fukushima, K. (1980). Neocognitron: A self-organizing neural network model for a mechanism of pattern recognition unaffected by shift in position, BioL Cybem. 36 (1980) 193-202.

Galway-Witham, J., & Stringer, C. (2018). How did Homo sapiens evolve? Science, 360(6395), 1296-1298.

Gandomi, A. H. (2014). Interior search algorithm (ISA): a novel approach for global optimization. ISA transactions, 53(4), 1168-1183.

Garcia, F. D., & Jacobs, B. (2011). Privacy-friendly energy-metering via homomorphic encryption. In International Workshop on Security and Trust Management (pp. 226-238). Springer, Berlin, Heidelberg.

Gaudreau, P., Miranda, D., & Gareau, A. (2014). Canadian university students in wireless classrooms: What do they do on their laptops and does it really matter? Computers & Education, 70, 245-255.

Geem, Z. W. (2010). Survival of the fittest algorithm or the novelest algorithm? The existence reason of the harmony search algorithm. International Journal of Applied Metaheuristic Computing, 1(4), 76-80.

Geem, Z. W., Kim, J. H., & Loganathan, G. V. (2001). A new heuristic optimization algorithm: harmony search. Simulation, 76(2), 60-68.

Goodfellow, I., Bengio, Y., Courville, A., & Bengio, Y. (2016). *Deep learning (Vol. 1).* Cambridge: MIT press.

Graeber, D. (2013). On the phenomenon of bullshit jobs: A work rant. Strike Magazine, 3, 1-5.

Graeber, D., & Cerutti, A. (2018). *Bullshit jobs.* New York, NY: Simon & Schuster.

Grier, D. A. (2001). Human computers: the first pioneers of the information age. Endeavour, 25(1), 28-32.

Grossberg, S. (1980). How does a brain build a cognitive code? Psychological Review, Vol. 87, No. 1, pp 1-51.

Hansen, N., and Ostermeier, A. (2001). Completely derandomized self-adaptation in evolution strategies. Evolutionary computation, 9(2), 159–195.

Hart, P. E., Nilsson, N. J., & Raphael, B. (1968). A formal basis for the heuristic determination of minimum cost paths. IEEE transactions on Systems Science and Cybernetics, 4(2), 100-107.

Harvey, I. (2013). Perspectives on Artificial Intelligence: Three Ways to Be Smart. In SmartData (pp. 27-38). Springer, New York, NY.

Hatamlou, A. (2013). Black hole: A new heuristic optimization approach for data clustering. Information sciences, 222, 175-184.

Havens, J. (2016). *Heartificial Intelligence: Embracing Our Humanity to Maximize Machines.* TarcherPerigee.

Hebb, D. O. (1949). *The organization of behavior: A neuropsychological theory.* Wiley, New York.

Helbing, D., Frey, B. S., Gigerenzer, G., Hafen, E., Hagner, M., Hofstetter, Y., ... & Zwitter, A. (2017). Digitale Demokratie statt Datendiktatur. In Unsere digitale Zukunft (pp. 3-21). Springer, Berlin, Heidelberg.

Hendler, J., & Mulvehill, A. (2016). *Social Machines: The Coming Collision of Artificial Intelligence. Social Networking, and Humanity.* Apress.

Hirose, M., & Ogawa, K. (2007). Honda humanoid robots development. Philosophical Transactions of the Royal Society of London A: Mathematical, Physical and Engineering Sciences, 365 (1850), 11–19.

Hoffmann, D. L., Standish, C. D., García-Diez, M., Pettitt, P. B., Milton, J. A., Zilhão, J., ... & Lorblanchet, M. (2018). U-Th dating of carbonate crusts reveals Neandertal origin of Iberian cave art. Science, 359(6378), 912-915.

Holland, J. H. (1975). *Adaptation in natural and artificial systems.* Ann Arbor, MI: University of Michigan Press.

Hopfield, J. J. (1982). Neural networks and physical systems with emergent collective computational abilities. Proceedings of the national academy of sciences, 79(8), 2554-2558.

Jaynes, J. (1976) *The origin of consciousness in the breakdown of the bicameral mind.* Houghton Mifflin.

Judd, J. S. (1990). *Neural network design and the complexity of learning.* MIT press.

Keenan, T. P. (2014). *Technocreep: the surrender of privacy and the capitalization of intimacy.* Greystone Books Ltd.

Kennedy, J. & Eberhart, R. (1995). Particle swarm optimization. Proc. IEEE International Conf. on Neural Networks (Perth, Australia), IEEE Service Center, Piscataway, NJ, 1995

Keynes J.M. (1930), "Economic Possibilities for our Grandchildren," in Essays in Persuasion (New York: Harcourt Brace, 1932), 358-373

Kirkpatrick, S., Gelatt, C. D., & Vecchi, M. P. (1983). Optimization by simulated annealing. Science, 220(4598), 671-680.

Kitano, H., & Asada, M. (1998). RoboCup humanoid challenge: That's one small step for a robot, one giant leap for mankind. In Proceedings of the 1998 IEEE/RSJ International Conference on Intelligent Robots and Systems (Vol. 1, pp. 419–424).

Kitano, H., Asada, M., Kuniyoshi, Y., Noda, I., Osawai, E., & Matsubara, H. (1998). Robocup: A challenge problem for AI and robotics. In RoboCup-97: Robot soccer world cup I (pp. 1–19). Berlin, Heidelberg: Springer.

Klein, J., & Takahata, N. (2002). *Where do we come from? The molecular evidence for human descent.* Springer Science & Business Media.

Klein, R. G. (2013). Modern human origins. General Anthropology, 20(1), 1–4.

Knuth, D. (1981). *The Art of Computer Programming, Second edition.* Addison Wesley Publishing Company, 1981

Kohonen, T. (1982). Self-organized formation of topologically correct feature maps. Biological cybernetics, 43(1), 59-69.

Kolmogorov, A. N. (1957). On the representation of continuous functions of many variables by superposition of continuous functions of one variable and addition. In Dokl. Akad. Nauk. SSSR, vol. 114, pp. 953-956.

Korf, R. E. (1985). Depth-first iterative-deepening: An optimal admissible tree search. Artificial intelligence, 27(1), 97-109.

Korf, R. E., & Zhang, W. (2000, July). Divide-and-conquer frontier search applied to optimal sequence alignment. In AAAI/IAAI (pp. 910-916).

Koza, J. R. (1992). *Genetic programming: on the programming of computers by means of natural selection.* MIT Press, Cambridge.

Kurzweil, R. (2005). *The singularity is near. When humans transcend biology,* New York, Viking.

Lepp, A., Barkley, J. E., & Karpinski, A. C. (2015). The relationship between cell phone use and academic performance in a sample of US college students. Sage Open, 5(1), 2158244015573169.

Li, X., Shi, M., & Wang, X. S. (2019). Video mining: Measuring visual information using automatic methods. International Journal of Research in Marketing, 36(2), 216-231.

Lipson, H., & Kurman, M. (2016). *Driverless: intelligent cars and the road ahead.* MIT Press.

Longo, B. (2015). *Edmund Berkeley and the Social Responsibility of Computer Professionals.* Morgan & Claypool Publishers.

Luger, G. F. (2009). *Artificial intelligence: structures and strategies for complex problem solving.* Pearson education.

Manyika, J., Chui, M., Brown, B., Bughin, J., Dobbs, R., Roxburgh, C., & Byers, A. H. (2011). Big data: The next frontier for innovation, competition, and productivity.

Matthias, A. (2011). Is the concept of an ethical governor philosophically sound. Tilting Perspectives: Technologies on the Stand: Legal and Ethical Questions in Neuroscience and Robotics, Tilburg University, The Netherlands.

May, L., Rovie, E., and Viner, S. (2005) The Morality of War: Classical and Contemporary Readings,Pearson-Prentice Hall.

McBrearty, S., & Brooks, A. S. (2000). The revolution that wasn't: a new interpretation of the origin of modern human behavior. Journal of human evolution, 39(5), 453-563.

McCarthy, J., Minsky, M., Rochester, N., & Shannon, C. (1955). A Proposal for the Dartmouth Summer Research Project on Artificial Intelligence. Formal Reasoning Group, Stanford University, Stanford, CA.

McCorduck, P. (1979). *Machines who think: A personal inquiry into the history and prospects of artificial intelligence.* San Francisco: Freeman, New York, NY, 1979.

McCulloch, W. S., & Pitts, W. (1943). A logical calculus of the ideas immanent in nervous activity. The bulletin of mathematical biophysics, 5(4), 115-133.

Mendel, G. (1866). Versuche uber pflanzen-hybriden. Verhandlungen des naturforschenden Vereins in Brunn fur, 4, 3-47.

Meng, X., Liu, Y., Gao, X., & Zhang, H. (2014). A new bio-inspired algorithm: chicken swarm optimization. In International conference in swarm intelligence (pp. 86-94). Springer, Cham.

Metz, C. (2016).Google's AI Wins Pivotal Second Game in Match With Go Grandmaster. WIRED. 10 March 2016

Michie, D., & Chambers, R. A. (1968). BOXES: An experiment in adaptive control. Machine intelligence, 2(2), 137-152.

Milgram, S., & van Gasteren, L. (1974). Das Milgram-Experiment. Rowohlt.

Millington, I., & Funge, J. (2009). *Artificial intelligence for games.* CRC Press.

Minsky, M. (1954). Theory of neural-analog reinforcement systems and its application to the brain-model problem. Princeton University (Ph.D dissertation).

Minsky, M. (1961). Steps toward artificial intelligence. Proceedings of the IRE, 49(1), 8-30.

Minsky, M. (1967). *Computation: finite and infinite machines.* Prentice-Hall, Inc.

Minsky, M. (1970) Life Magazine, Nov 20th, 1970, p. 68

Minsky, M., & Papert, S. (1969). *Perceptrons: An Introduction to Computational Geometry.* MIT press

Minsky, M., & Papert, S. (1988). *Perceptrons* (expanded edition). MIT Press, Cambridge, MA.

Mirjalili, S., & Lewis, A. (2016). The whale optimization algorithm. Advances in Engineering Software, 95, 51-67.

Mitchell, T. (1997). *Machine Learning.* MacGraw-Hill Companies. Inc.

Mori, M. (1970). Bukimi no tani (the Uncanny Valley). Energy, 7, 33–35.

Mueller, Pam A., and Daniel M. Oppenheimer. "The pen is mightier than the keyboard: Advantages of longhand over laptop note taking." Psychological science 25, no. 6 (2014): 1159-1168.

Nagel, T. (1974). What is it like to be a bat? The philosophical review, 83(4), 435-450.

Najafabadi, M. K., Mohamed, A. H., & Mahrin, M. N. R. (2019). A survey on data mining techniques in recommender systems. Soft Computing, 23(2), 627-654.

Nelson, A., Barlow, G., and Doitsidis, L. (2009). Fitness functions in evolutionary robotics: A survey and analysis. Robotics and Autonomous Systems, 57(4):345–370.

Newell, A., & Ernst, G. (1965). The search for generality. In Proc. IFIP Congress (Vol. 65, pp. 17-24).

Newman, J. H. (1852) The Idea of a University. London: Aeterna Press, 2015.

Noble, D. F. (1998). Digital diploma mills, part 1: The automation of higher education. October, 86, 107-117.

Nolfi, S., & Floreano, D. (2000). *Evolutionary robotics. The biology, intelligence, and technology of self-organizing machines.* Cambridge: MIT Press.

Norman, D.A. (1993). *Things that make us smart.* Reading, MA: Addison Wesley.

O'Donohue, J. (1997) *Anam Cara: Spiritual Wisdom from the Celtic World,* Bantam Press.

O'Neill M (2001) Automatic Programming in an Arbitrary Language: Evolving Programs with Grammatical Evolution. PhD thesis. University of Limerick, Ireland

O'Connell, M. (2018). *To be a machine: Adventures among cyborgs, utopians, hackers, and the futurists solving the modest problem of death.* Anchor.

Odili, J. B., & Kahar, M. N. (2016). Solving the traveling Salesman's problem using the African Buffalo optimization. Computational intelligence and neuroscience, 2016, 3.

Odili, J. B., Kahar, M. N. M., & Anwar, S. (2015). African buffalo optimization: a swarm-intelligence technique. Procedia Computer Science, 76, 443-448.

Orwell, G. (1933). Down and Out in Paris and London (London: Gollancz).

Orwell, George, (1949). *Nineteen Eighty-Four*, 1990 ed. Penguin, London.

Ozawa, A., Furusato, R., & Yoshida, Y. (2016). Determining the relationship between a household's lifestyle and its electricity consumption in Japan by analyzing measured electric load profiles. Energy and Buildings, 119, 200-210.

Pan, J. S., Dao, T. K., Kuo, M. Y., & Horng, M. F. (2014). Hybrid bat algorithm with artificial bee colony. In Intelligent Data analysis and its Applications, Volume II (pp. 45-55). Springer, Cham.

Patterson, R. W., & Patterson, R. M. (2017). Computers and productivity: Evidence from laptop use in the college classroom. Economics of Education Review, 57, 66-79.

Pinker, S. (2018). *Enlightenment Now: The Case for Reason, Science, Humanism, and Progress.* Penguin.

Porambage, P., Ylianttila, M., Schmitt, C., Kumar, P., Gurtov, A., & Vasilakos, A. V. (2016). The quest for privacy in the internet of things. IEEE Cloud Computing, 3(2), 36-45.

Postman, N. (1985), *Amusing Ourselves to Death: Public Discourse in the Age of Show Business,* Viking, New York, NY.

Quinlan, J. R. (1986). Induction of decision trees. Machine learning, 1(1), 81-106.

Quinlan, J. R. (1993*). C4.5: Programs for machine learning.* Elsevier.

Rabin, S. (Ed.). (2002). *AI game programming wisdom.* Charles River Media.

Rabin, S. (Ed.). (2004). *AI Game Programming Wisdom 2.* Charles River Media.

Rabin, S. (Ed.). (2006). *AI Game Programming Wisdom 3.* Charles River Media.

Rabin, S. (Ed.). (2014). *AI Game programming wisdom 4.* Charles River Media.

Rechenberg, I. (1973) Evolutionsstrategie: Optimierung technischer Systeme nach Prinzipien der biologischen Evolution. Frommann-Holzboog Verlag, Stuttgart

Rich, E. (1983). *Artificial Intelligence.* McGraw-Hill series in artificial intelligence.

Rid, T. (2016). *Rise of the machines: A cybernetic history.* WW Norton & Company.

Rochester, N., Holland, J., Haibt, L., & Duda, W. (1956). Tests on a cell assembly theory of the action of the brain, using a large digital computer. IRE Transactions on information Theory, 2 (3), 80-93.

Rosenblatt, F. (1958). The perceptron: a probabilistic model for information storage and organization in the brain. Psychological review, 65(6), 386.

Rumelhart, D. E., Hinton, G. E., & Williams, R. J. (1986). Learning representations by back-propagating errors. nature, 323(6088), 533.

Rumelhart, D. E., McClelland, J. L., & PDP Research Group. (1986). *Parallel Distributed Processing, Vol. 1. and Vol. 2.* MIT press, Cambridge MA.

Russell, S., & Norvig, P. (2010). *Artificial Intelligence: A modern approach.* 3rd Edition. Pearson education.

Samuel, A. L. (1959). Some Studies in Machine Learning Using the Game of Checkers. IBM Journal of Research and Development, 3(3), 210–229.

Sarkar S. & Pfeifer J.(eds.) (2006) *Philosophy of Science –An Encyclopaedia.* Routledge.

Schrittwieser, J., Antonoglou, I., Hubert, T., Simonyan, K., Sifre, L., Schmitt, S., ... & Lillicrap, T. (2019). Mastering atari, go, chess and shogi by planning with a learned model. arXiv preprint arXiv:1911.08265.

Schwyzer, H. (1973). Thought and reality: the metaphysics of Kant and Wittgenstein. The Philosophical Quarterly (1950-), 23(92), 193-206.

Sejnowski, T. J., and Rosenberg, C.R. (1986). NETtalk: A parallel network that learns to read aloud. The Johns Hopkins University EE and CS Tech. Rep. IHU/EECS-86/01.

Shadbolt, N., & Hampson, R. (2019). *The digital ape: how to live (in peace) with smart machines.* Oxford University Press.

Shannon, C. E. (1938). A symbolic analysis of relay and switching circuits. Electrical Engineering, 57(12), 713-723.

Shannon, C. E. (1948). A mathematical theory of communication. Bell system technical journal, 27 (3), 379-423.

Shannon, C. E. (1950). Programming a computer for playing chess. The London, Edinburgh

Silver, D., Huang, A., Maddison, C. J., Guez, A., Sifre, L., Van Den Driessche, G., ... & Dieleman, S. (2016). Mastering the game of Go with deep neural networks and tree search. nature, 529 (7587), 484-489.

Silver, D., Hubert, T., Schrittwieser, J., Antonoglou, I., Lai, M., Guez, A., Lanctot, M., Sifre, L., Kumaran, D., Graepel, T., Lillicrap, T., Simonyan, K., & Hassabis, D. (2017). Mastering Chess and Shogi by Self-Play with a General Reinforcement Learning Algorithm. arXiv preprint arXiv:1712.01815.

Silver, D., Hubert, T., Schrittwieser, J., Antonoglou, I., Lai, M., Guez, A., Lanctot, M., Sifre, L., Kumaran, D., Graepel, T. and Lillicrap, T. (2018). A general reinforcement learning algorithm that masters chess, shogi, and Go through self-play. Science, 362(6419), 1140-1144.

Silver, D., Schrittwieser, J., Simonyan, K., Antonoglou, I., Huang, A., Guez, A., ... & Chen, Y. (2017b). Mastering the game of go without human knowledge. Nature, 550(7676), 354.

Sims, K. (1994a). Evolving virtual creatures. In Proceedings of the 21st Annual Conference on Computer Graphics and Interactive Techniques (pp. 15–22). ACM.

Sims, K. (1994b). Evolving 3D morphology and behavior by competition. In R. Brooks, & P. Maes, (Eds.), Proceedings of artificial life IV (pp. 28–39). Cambridge, MA: MIT Press.

Singh, B., & Singh, H. K. (2010). Web data mining research: a survey. In 2010 IEEE International Conference on Computational Intelligence and Computing Research (pp. 1-10). IEEE.

Sipper, M. (2002). *Machine Nature. The Coming Age of Bio-Inspired Computing*. McGraw-Hill.

Sipper, M., Sanchez, E., Mange, D., Tomassini, M., Pérez-Uribe, A., & Stauffer, A. (1997). A phylogenetic, ontogenetic, and epigenetic view of bio-inspired hardware systems. IEEE Transactions on Evolutionary Computation, 1(1), 83-97.

Slagle, James R. "A heuristic program that solves symbolic integration problems in freshman calculus." Journal of the ACM (JACM) 10.4 (1963): 507-520.

Slocum, J., & Sonneveld, D. (2006). The 15 Puzzle: How it Drove the World Crazy: the Puzzle that Started the Craze of 1880; how Amercia's Greatest Puzzle Designer, Sam Loyd, Fooled Everyone for 115 Years. Slocum Puzzle Foundation.

Sörensen, K. (2015). Metaheuristics—the metaphor exposed. International Transactions in Operational Research, 22(1), 3-18.

Stanley, K. O., & Lehman, J. (2015). *Why greatness cannot be planned: The myth of the objective*. Springer.

Stanley, K. O., and Miikkulainen, R. (2002). Evolving neural networks through augmenting to-pologies. Evolutionary computation, 10(2), 99–127.

Steinerberger, S. (2015). On the number of positions in chess without promotion. International Journal of Game Theory, 44(3), 761-767.

Stentz, A. (1994). Optimal and efficient path planning for partially-known environments. In Robotics and Automation, 1994. Proceedings., 1994 IEEE International Conference on (pp. 3310-3317). IEEE.

Sutton, R. S., & Barto, A. G. (1998). Introduction to reinforcement learning (Vol. 135). Cambridge: MIT press.

Taherdangkoo, M., Shirzadi, M. H., Yazdi, M., & Bagheri, M. H. (2013). A robust clustering method based on blind, naked mole-rats (BNMR) algorithm. Swarm and Evolutionary Computation, 10, 1-11.

Takagi, H. (2001). Interactive evolutionary computation: Fusion of the capabilities of EC optimization and human evaluation. Proceedings of the IEEE, 89(9), 1275–1296.

Teeuwen, M. (2002). From Jindo to Shinto: A concept takes shape. Japanese Journal of Religious Studies, 233-263.

Tegmark, M. (2017). *Life 3.0: Being Human in the Age of Artificial Intelligence*. Allen Lane.

Teodorovic, D., Lucic, P., Markovic, G., & Dell'Orco, M. (2006, September). Bee colony optimization: principles and applications. In Neural Network Applications in Electrical Engineering, 2006. NEUREL 2006. 8th Seminar on (pp. 151-156). IEEE.

Tian, Y., Ma, J., Gong, Q., Sengupta, S., Chen, Z., Pinkerton, J., & Zitnick, C. L. (2019). Elf opengo: An analysis and open reimplementation of alphazero. arXiv preprint arXiv:1902.04522.

Tsai, P. W., & Istanda, V. (2013). Review on cat swarm optimization algorithms. In Consumer Electronics, Communications and Networks (CECNet), 2013 3rd International Conference on (pp. 564-567). IEEE.

Tsai, P. W., Pan, J. S., Chen, S. M., Liao, B. Y., & Hao, S. P. (2008, July). Parallel cat swarm optimization. In Machine learning and cybernetics, 2008 international conference on (Vol. 6, pp. 3328-3333). IEEE.

Turing, A. M. (1950). Computing machinery and intelligence. Mind, 59, 433–460.

Turing, A. M. (1952). Transcript from a 1952 radio broadcast. Available from the Alan Turing Archives.

Turkle, S. (2015). *Reclaiming conversation: The power of talk in a digital age.* Penguin.

Valente, J., Wynn, M. A., & Cardenas, A. A. (2019). Stealing, Spying, and Abusing: Consequences of Attacks on Internet of Things Devices. IEEE Security & Privacy, 17(5), 10-21.

von Neumann, J. (1928): "Zur Theorie der Gesellschaftsspiele." Mathematische Annalen, 100, 1928, pp. 295–320.

von Neumann, J. (1955). 'Can We Survive Technology?', Fortune, June issue.

von Schiller, F. (1795). Über die ästhetische Erziehung des Menschen in einer Reihe von Briefen.

Wang, M., & Deng, W. (2018). Deep face recognition: A survey. arXiv preprint arXiv:1804.06655.

Wang, Y., & Kosinski, M. (2018). Deep neural networks are more accurate than humans at detecting sexual orientation from facial images. Journal of personality and social psychology, 114(2), 246.

Webb, G. (5th February 2015) Say Goodbye to Privacy. WIRED editorial.

Werbos, P. (1974). Beyond regression: new fools for prediction and analysis in the behavioral sciences. PhD thesis, Harvard University.

Weyland, D. (2010). A rigorous analysis of the harmony search algorithm: How the research community can be misled by a "novel" methodology. International Journal of Applied Metaheuristic Computing (IJAMC), 1(2), 50-60.

Weyland, D. (2015). A critical analysis of the harmony search algorithm—How not to solve sudoku. Operations Research Perspectives, 2, 97-105.

Widrow, B., & Hoff, M. E. (1960). Adaptive switching circuits (No. TR-1553-1). Stanford Univ. CA Stanford Electronics Labs.

Wiener, N. (1954*). The Human Use of Human Beings.* Houghton Mifflin Company, New York.

Wilkes, M. V. (1977). Babbage as a computer pioneer. Historia Mathematica, 4(4), 415-440.

Witten, I. H., Frank, E., Hall, M. A., & Pal, C. J. (2016). *Data Mining: Practical machine learning tools and techniques.* Morgan Kaufmann.

Wittgenstein, L. (2009). *Philosophical investigations.* John Wiley & Sons.

Wittgenstein, L., Anscombe, G. E. M., & Rhees, R. (1953). *Philosophische Untersuchungen.* (Philosophical investigations).

Yang, X. S. (2009). Harmony search as a metaheuristic algorithm. In Music-inspired harmony search algorithm (pp. 1-14). Springer, Berlin, Heidelberg.

Yoshiike, T., Kuroda, M., Ujino, R., Kaneko, H., Higuchi, H., Iwasaki, S., ... & Koshiishi, T. (2017). Development of experimental legged robot for inspection and disaster response in plants. In 2017 IEEE/RSJ International Conference on Intelligent Robots and Systems (IROS) (pp. 4869-4876). IEEE.

Zimmer, C. (2001). *Evolution: the triumph of an idea.* Random House.

Zurada, J. M. (1992). *Introduction to artificial neural systems.* St. Paul: West.

Printed in the United States
by Baker & Taylor Publisher Services